# 5-minute Memory Workout

Sean Callery

HarperCollins Publishers Ltd
77–85 Fulham Palace Road
London
W6 8JB

**www.collins.co.uk**

Collins is a registered trademark of HarperCollins Publishers Ltd.

First published in 2007
Text © 2007 HarperCollins Publishers
Illustrations © 2007 HarperCollins Publishers

10 09 08 07
7 6 5 4 3 2 1

The author asserts his moral right to be identified as the author of this work. All rights reserved. No parts of this publication may be reproduced, stored in a retrieval system or transmitted, in any form or by any means without the prior permission of the publishers.
A catalogue record for this book is available from the British Library.

ISBN-10:0-00-725121-1
ISBN-13:978-0-00-725121-6

Designed by Martin Brown
Printed and bound in Italy by Amadeus

# CONTENTS

| | |
|---|---|
| **INTRODUCTION** | 4 |
| **KEEPING YOUR BRAIN FIT** | 15 |
| **WAKING YOUR BRAIN UP** | 75 |
| **SHORT CUTS** | 113 |
| **SPRINTING** | 141 |
| **GLOSSARY** | 183 |
| **FURTHER INFORMATION** | 187 |
| **INDEX** | 190 |

# INTRODUCTION

Memory is just one function of our extraordinary brain. This mighty organ runs our bodies – itself an amazing feat – enables us to feel emotions, to think and dream and is a storehouse of information about everything we have experienced, real or imaginary, which allows us to build knowledge and understanding throughout our lives.

The brain is our heaviest (1-1.5 kg), hungriest (it burns up nearly a quarter of our energy) and most complex organ, more powerful than any computer in the range of processes it can undertake, including, of course, memory. Yet we barely understand how it works.

### 'We only use 10 per cent of our brain'

No we don't, but this is a persistent myth often quoted to suggest the brain has vast untapped powers. Brain scans show activity across the whole of the brain even when we are asleep, and we are affected if any part of the brain is damaged – showing that we are not harbouring a 90 per cent unused instrument. The myth may originate from research suggesting that only one in ten neurons are active at any one time.

# Command and control

The brain is the control and command centre of the central nervous system. Our brains have more than 100 billion neurons, or electrically active nerve cells, known as grey matter. A trillion or so glial cells, known as white matter, help each neuron connect with up to 10,000 of its fellow cells, forming a web of links that are crucial for an effective memory.

The human brain has evolved over many thousands of years into a complex machine. It developed from and still has at its centre (at the base of the skull) the reptilian brain, or brain stem, which controls life functions and influences whether we will fight, hide or run when we are feeling frightened or angry.

The phrase 'rush of blood to the head' describes the way our inner reptilian brain can dominate during times of high emotion. For example, some people react to confrontations by becoming aggressive, others freeze like a rabbit stuck in the headlights of a car, while others flee.

Those who respond calmly have trained themselves to curb the urges of this impulse-led brain and employ the many layers of more sophisticated reasoning that have been added to this foundation. This shows that we can train and control our brains, a capability that can also apply to our memory functions.

The brain is divided into two hemispheres. In broad terms, the right side deals with emotions and the left side deals with rational thought. Each hemisphere also has four lobes: the frontal lobe, the temporal lobe, the parietal lobe and the occipital lobe. The frontal and temporal lobes are most involved with memory functions.

We rely on our memory in all aspects of life, from when and how to brush our teeth to running meetings and telling bedtime stories. We are not usually conscious that this is happening. Much of the memory function is automatic, and it isn't like a muscle that you can train to perform better. However, by changing the way that we think, by learning memory techniques that exploit the brain's interconnections, and by working at how to remember rather than just hoping it will happen, we can feel more in control and that will improve our memory function.

Self-knowledge and emotional intelligence (understanding our own emotions and, from this, the feelings of others) make us happier and calmer and better able to comprehend concepts, while we can also train ourselves to learn better with efficient study techniques and sustained concentration. Anyone can train themselves to have better understanding and recall of information, which in turn improves their quality of life.

# How memory works

There isn't a single part of the brain dedicated to dealing with memory, and scientists don't have a full understanding of how memory operates. However, its workings are best described in three stages, and we can refine each of these to work more effectively for us.

### Stage 1
This is encoding, when we take in information. We don't remember everything that happens to us – we'd go barmy if we did. We can remember about seven things for roughly 30-40 seconds in our short-term, or working, memory. After that, these thoughts are mostly filtered out (you might consider the phrase 'in one ear and out the other' appropriate), while we retain what we regard as the useful bits at the time. Making a conscious effort to remember things, and devising and using mental filing systems helps us to remember what we need to remember and to block out what we don't.

### Stage 2
This is when we store information permanently. There are many ways in which this happens, but the more associations, or links, the thought has with other thoughts, the more likely it is to stay in our memory. This book shows you how to create

associations and how to deliberately link thoughts together into a chain, making it easier to move to stage 3.

## Stage 3
This is recalling. We've all had that feeling of consciously trying to remember something, failing, then having it pop into our head a few minutes later. This is because the signals from neurons are zooming around 'looking' in mental filing cabinets and seeking out connections to what we want to know. They will do this faster if we created a reliable 'route' (or more likely a series of 'routes') in the first place. We can train ourselves to do this and thus retrieve information efficiently, so that we are less likely to spend several minutes agonising over our failure to remember something then find it popping into our head a few minutes after we really needed it.

## Haven't you read this already?
That odd feeling that you've come across something before, known as 'déjà vu', seems to be caused by an overactive memory function which creates a memory as it is processing the present, giving the false impression that the experience is being lived through for a second time.

## The short-term filter

The short-term or working memory processes information about things that have just happened, filtering what can be ignored and what needs to be retained in the long-term memory. It takes in information from the senses (iconic memory for sight, echoic memory for sounds, haptic memory for touch) and can typically hold on to about seven items for roughly 30-40 seconds, after which the memories fade and are replaced by new input. That's why you can remember a phone number long enough to dial it, but then lose the ability to recall it.

## What makes it to my long-term memory?

Long-term memory is generally divided into three broad roles:

### Semantic memory

We store our memory of the world in the semantic memory. This is the kind of memory you use in a pub quiz as it holds facts, rules, meanings and concepts such as social customs. It is the brain's equivalent of an encyclopedia and does not involve the senses. While other memory functions typically start to deteriorate when we are in our thirties, semantic memory survives well.

### Procedural memory
Procedural skills are those such as driving, swimming or hitting a ball that become automatic and unconscious.

### Episodic memory
Episodic memory is where you store the biography of your life so far. What did you have for breakfast today? What happened on your last birthday? This type of information, which is linked to time, is the memory of incidents in our lives, including how they affected our senses.

However, these definitions are ones we have imposed on the brain: it isn't organised into separate distinct departments in this neat way. We could continue creating sub-sections – for example, to include 'declarative memory', the things you decide to remember, like your own phone number – or 'prospective memory': the list of future actions, from turning off the cooker to catching a plane. Memory is multi-layered and multi-sensory, and highly sophisticated. It doesn't always do what we want it to, but it can be trained to do some things better.

## Memory curves
Much of how we define different elements of memory stems from the work of German psychologist Hermann Ebbinghaus (1850–1909), who devised new ways to test and analyse memory.

He was the first to measure how long we retain memories, known as the forgetting curve, and how we take in and adopt fresh information, known as the learning curve.

## I've started forgetting things

As we age, our bodies change and with this we tend to lose mental sharpness: most people's memories start to worsen marginally from their thirties, and function significantly less well after the age of about 65. Of course, you could argue that this is because they have more memories cluttering up their brains by then. Forgotten information either never made it through the short-term memory filter, or has since got 'lost' in the long-term memory, so that now you can't find it. However, you can counteract the loss of recall ability.

A research programme investigated how practising mental sharpness can improve memory. Older people who volunteered for the programme found that their recollecting abilities actually improved when they had been trained in memory techniques.

## It's on the tip of my tongue

That frustrating feeling that a word is somewhere in your head just out of your mental reach is very common and has been tormenting us for thousands of years (the ancient Greek writer Aristotle refers to it). The 'tip of the tongue phenomenon' worsens with age and is particularly prevalent (and embarrassing) when it affects our ability to recall names. Many psychologists have studied the phenomenon, and posited various theories about why and how it happens. What seems clear is that the more you train your memory to store information efficiently so that you can retrieve it through a variety of routes, the less likely you may be to suffer this irritation.

## The gender gap

Men have about 100g more brain tissue on average than women because men are generally bigger: the sexes have similar brain-weight-to-body-weight ratios. This does not mean men are cleverer: there is no link between human brain size and intelligence. Woman's brains seem to have a stronger link between the left and right hemispheres because they have a slightly denser corpus callosum (the connecting structure), which means the rational and emotional sides might communicate more effectively and may explain why women multi-task better.

Pregnant women often complain that their memory suffers during pregnancy. This temporary problem is probably caused by a combination of psychological, hormonal and chemical factors.

# How to use this book

The memory techniques in this book are ordered in increasing degrees of complexity, starting with relatively simple methods and building up to the practices used by world memory champions who have honed their abilities over many years. It is worth trying every exercise as this will clarify how each technique works and give you a feel for whether it is right for you. It will also be interesting for you to return to the initial assessment and awareness exercises (see pages 16–21) and see how your performance improves as you learn new memory skills. Repeat exercises for practice of those particular skills that interest you – memory champions develop their abilities over time.

## How memorable is this?

We are besieged with sales information, most of which we ignore and forget pretty quickly (putting it in the bin both physically and mentally). In the marketing world, the skill of making something memorable is like gold dust. In California, a machine is being developed that may be capable of telling what you remember now and in the future. Using a technique called functional brain imaging (fMRI), it places people in a giant magnet and measures their brain responses to various stimuli. By concentrating on the parts of the brain where long-term memories are encoded, it may be able to predict what will remain in our memories, and what will be ignored.

### Raise my IQ

One way of measuring intelligence is an IQ (intelligence quotient) test, which sets a number of verbal and non-verbal reasoning tasks to produce a standardised score. It was thought that our IQ was set genetically and could not be changed, but new research suggests we can raise our scores by 8 per cent after training our brains. The key factor is improving working memory – the brain's short-term information storage system – in this case through tasks such as memorising the positions of dots on a grid. Importantly, the improvements discovered were not just apparent in carrying out IQ test-type activities, but other cognitive skills too.

# KEEPING YOUR BRAIN FIT

## Introduction

A fit brain is open to new ideas and quick to learn, so let's take your mind to the gym. This section has information on what you can do to get and keep your brain in tip-top condition. This includes physical and mental activities that will help you to upgrade the performance of your memory.

There is also guidance on how to find your preferred methods of learning, which strongly influence how and what you remember. We explore the different senses and sensations that affect memory – memory is multi-sensory – to help you reach a better understanding of how your brain operates and therefore help you find routes towards improving your memory.

## Memory check

You want to improve your memory – that's why you are reading this book. But how well is your memory working at the moment? These exercises will give you an idea of how efficiently you are storing information. You may be surprised at how well, or badly, you do this. Return to these exercises regularly so that you can measure the improvement as you start to use the techniques described later in the book.

## What am I good at remembering?

Few people have a poor memory for everything, but many of us find there are certain things that never make it into our memory. The following activities aim to help you identify what you are able to recall well, and where your memory lets you down.

### Life story

This is a test of episodic memory, drawing on what events from the last year have been retained in your long-term memory. Answer these:

1 What was the last DVD you bought or rented?
2 What did you do on your last birthday?
3 Who was with you on Christmas day?
4 What's the most recent TV programme you saw?
5 What colour was the last dog you saw?
6 What did you eat last Saturday night?
7 How did you celebrate the new year?
8 What was the most recent sporting event you witnessed and what was the outcome?
9 When was the last time you bought a pair of shoes?
10 Who got married at the last wedding you went to?
11 How did you celebrate your 18th birthday?
12 When did you last see your father?

A score of less than 7 is poor, and an average score would be 7-10 items.

## Detail recall

How much can you recall from reading a text? Read the story in the box carefully, but without taking notes.

Ashran's watch said **8.51** as he placed his **black case** next to his **crocodile skin shoes** and fumbled in his **trenchcoat pocket** for his **plane ticket** to Paris. It was long enough for **Baz** to seize the **leather handle** and sneak the case under his **blue jacket**. He scurried to the **toilets** to check his find.

Behind the locked cubicle door he groaned when he saw the **two gold bars** and four **purple velvet bags** of diamonds and rubies. This meant danger: the owner would **kill** to get this lot back. Gently closing the orange door behind him, he left his loot **on the toilet seat** and headed for the **bus stop**. In the next cubicle, **Ashran** reluctantly keyed in the last three digits of his **boss's number: 947**.

'**Sheila**, I missed my flight. Everything's booked solid until **Thursday**,' he lied.

'No problem,' came the reply. 'I'll come to you. Meet me at the **concourse newsagent** in **four hours**.'

He barely heard the **startled cry** from the **next door booth**, or the heavy footsteps running away... Then see overleaf.

## Detail recall, part II

Close the book and write down as many details from it as you can. Check to see how many out of the 28 emboldened details you remembered.

An average score is between 16 and 21. Less is obviously poorer, more is impressive.

## Short-term numbers

This is a test of your short-term memory. Get someone to read you a sequence of five single-digit numbers. You should be able to repeat the numbers back quite easily. Ask them to provide another five digits, which this time you must say in reverse order.

Continue these two activities with an extra digit added each cycle until you make a mistake. This uses your short-term memory, which can store between 5 and 9 items. Reversing the digit order forces you to manipulate the information coming in.

# Where was it?

This is a test of your visual and spatial memory.

| | | | | |
|---|---|---|---|---|
| Row A | ● | ● | ▲ | ■ |
| Row B | ■ | ✱ | ✱ | ● |
| Row C | ■ | ■ | ■ | ■ |
| Row D | ✱ | ▲ | ✱ | ✱ |

Study the table for one minute before answering these questions without referring to it.

1. How many squares are there?
2. How many circles are there?
3. How many triangles are there?
4. How many stars are there?
5. What is the shape in Row C?
6. What is below the orange triangle?
7. What is above the black square?
8. What is in the top right corner?
9. What is two above the triangle in Row D?
10. What is underneath the top square?
11. What is three below the orange triangle?
12. What is between the two black squares in the right hand column?

Check your answers (see page 186). A score of 6-9 is average. Below that is poor, above it shows very good visual and spatial memory.

## In my day

In the evening, try to recall everything you did today, in order. Start by remembering every detail, from what you had for breakfast to who you met first during working hours and what you ate with your morning drink. Edit this into a list of ten significant things you did (after all, the number of biscuits doesn't really matter, but attending a meeting or helping a friend does).

Now it's time to test your everyday memory by removing some key triggers.

## What happened before that?

After the following day, try scrolling through the day backwards so that you recall its events in reverse order. Check afterwards to see if there are any gaps where parts of the day are unaccounted for.

These two exercises test your recall over a period of several hours and, with the backwards recalling, your ability to control your memory and find triggers for recall. We tend to recall events chronologically, so scanning backwards is trickier. Identifying effective triggers is vital for improving our memory.

The next step is to start to identify what sort of things you forget, narrowing down where the weaknesses are in your memory and what factors could be causing them.

### Memory diary

 For one week, keep a lifestyle and memory diary. Answer these questions for each day:

- ✔ Did you wake up feeling refreshed?
- ✔ Did you have breakfast?
- ✔ What were the key events of the day?
- ✔ How good was your concentration at these?
- ✔ Did you forget anything important during the day?
- ✔ Did you use any memory aids (such as diary notes, stickies etc)?
- ✔ How stressed did you feel?
- ✔ Did you eat healthily?
- ✔ What was your alcohol intake?
- ✔ What time did you go to bed?

After a week you will begin to build a picture of how you are using your memory and whether your lifestyle is affecting its performance. Is there a pattern between poor sleep or high alcohol consumption and poor memory? What other factors might be influencing your ability to remember?

You may wish to repeat this exercise at a later stage once you are familiar with the techniques in this book.

### Memory and personality type

Another factor that can affect your ability to remember is your personality type. Fast-talking extroverts who are always on the move are likely to spend less time recalling names and other information than more introverted people who like to take their time. Someone who is a good listener probably talks less but pays more attention to what is said and enjoys spending time on their own, when they can go over the day's events and conversations. These types of people are likely to have developed better memory skills.

# GETTING YOUR BRAIN FIT: PHYSICAL ACTIVITY

Exercise is good for your body and your brain. Aerobic exercises (movements such as walking, running, cycling and swimming that increase the heart rate for a while) help our brains because they push up the heart rate, boosting the amount of oxygen and nutrients circulating in the blood around the whole body, including its heaviest organ.

You don't have to bust a gut in the gym: a gentle 30-minute walk three times a week has been found to improve our ability to learn, concentrate and reason by 15 per cent. One study has found that the slow, controlled stretches of yoga provide a mental lift (it is thought that exercises involving bending over backwards work particularly well).

Brain activity is reduced when we feel anxious, while exercise lowers our stress levels and has also been found to stimulate the growth of new brain cells. If you don't exercise much, start slowly and increase the amount of activity gradually. Walk or cycle short journeys rather than driving or taking the bus, and use the stairs instead of the lift.

## BRAIN-BOOSTING ACTION

Some doctors believe there are specific exercises that benefit parts of the brain by pressing on points of the body. This is similar to how acupuncture practitioners put needles into key points that are not necessarily close to the area they are treating. There is controversy about whether these exercises have genuine benefits (see page 28): the only way to judge is to try them for a while and decide if they work for you.

Try this general 'brain booster', which is claimed to increase blood flow and help concentration skills:

### Blow flow

**1** Stretch the thumb and index finger of one hand as far apart as possible.
**2** Place your index finger and thumb either side of the sternum (breastbone) in the slight indentations below the collar bone, and press regularly and lightly.
**3** While doing this, put your other hand over the navel and press gently.
**4** Do this for two minutes.

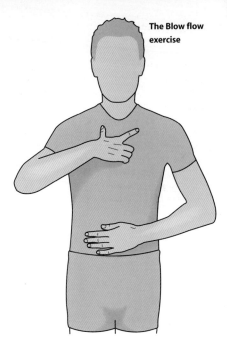

**The Blow flow exercise**

The next movement is designed to aid information flow between the left and right hemispheres of the brain.

### Left and right

1. Standing or sitting, gently raise the left knee, placing the right hand on it as you do so. Then lower it.
2. Repeat with the right knee and left hand.
3. Keep going for two minutes.

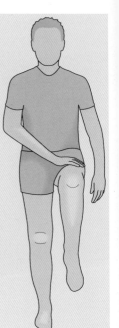

Concentration is crucial to a good memory, allowing you to absorb information quickly. This is a mind-calming exercise to improve this ability:

## Calming down

1 Stand or sit and cross the right ankle over the left one.
2 Cross the right wrist over the left wrist, linking the fingers.
3 Push your elbows out and turn your fingers towards your body to rest them on the sternum (breastbone). Hold this position.
4 Breathe evenly for several minutes.

## A stimulating argument

The theory behind these exercises is that we tend to 'switch off' one brain hemisphere when the other side is more actively engaged, reducing its efficiency and creating 'learning blocks'. Proponents believe that certain movements activate the mind and body system, stimulating parts of the brain which would otherwise be functioning less well.

While doctors agree that exercise in general is good for the brain, some say it is nonsense to suggest that we can stimulate parts of it by applying pressure at certain points. They point out that you cannot encourage blood flow along the interior of the body just by pushing down on the outside of its frame. As with all the exercises in this book, try them out a few times and decide what works for you – even if the science doesn't make sense, if you feel a benefit, it's working.

# EATING YOUR WAY TO A BETTER MEMORY

Diet can help your memory. First, our bodies function best when properly hydrated, so drink plenty of water – at least eight glasses a day. About 80 per cent of brain tissue (grey matter) is water, so even mild dehydration will have an effect on its performance. Follow guidance on healthy diets such as eating wholemeal and whole grain and unprocessed, unrefined foods. Eating a substantial breakfast helps you maintain concentration levels by kicking off the day with a good level of fuel in your tank.

Some foods have been particularly linked with suggested memory improvement.

- ✔ Blueberries and strawberries: these and other dark fruits help our short-term memory, coordination and concentration because they are high in the antioxidants that deal with free radicals (unstable molecules that react with oxygen and damage the brain).
- ✔ Broccoli: this contains substances that inhibit the enzyme that breaks down acetylcholine, one of the brain neurotransmitters (chemicals involved in the transmission of nerve impulses between nerve cells). These substances (found at lower

levels in other members of the cabbage family and also in yogurt and eggs) act in the same way as drugs that treat degenerative brain conditions.
- ✔ Carbohydrates: these are found in foods such as grains, fruit and vegetables, and are a source of the neurotransmitter serotonin.
- ✔ Curry: the spice turmeric may have a role in slowing down brain disease. It contains a compound called curcumin, which seems to reduce the build-up of tangles of cells in the brain known as amyloid plaques.
- ✔ Fibre: high-fibre foods such as beans and wholemeal bread have been shown to boost brain activity.
- ✔ Fish: the old saying that fish is brain food turns out to be true: oily fish such as mackerel, salmon, tuna and sardines are rich in Omega 3 oils. These are fatty acids that (among other things) maintain the fluidity of the membranes in the brain. Nuts and seeds are another source of these fatty acids.
- ✔ Ginkgo biloba: leaves from this ancient tree are said to improve brain function and slow the progression of Alzheimer's. The effect is stronger when it is combined with another herbal remedy, ginseng.
- ✔ Iron: we need iron to maintain concentration and energy. Iron-rich foods include red meat, leafy green vegetables, fortified breakfast cereals,

sardines and eggs. Iron supplements are also available.
- ✔ Lemon balm: this plant helps us to stay calm, and in tests has improved people's ability to retrieve information, increasing the activity of the chemical messenger acetylcholine.
- ✔ Sage: this ancient herb has aided word recall in tests. Herbalists have long sung its praises: for example, in 1597 John Gerard wrote that it was 'singularly good for the head and quickeneth the nerves and memory'.
- ✔ Salad: raw salad leaves are full of brain-maintaining antioxidants.
- ✔ Yeast extract: this is high in B vitamins, which helps the brain function. Spread yeast extract products, such as Marmite and Vegemite on your toast for a healthy dose of B vitamins every morning.

## Fat on the brain

Excess weight slows down your brain. Researchers found that while those with healthy body mass index (BMI) could recall 56 per cent of words in a vocabulary test, the figure fell to 44 per cent if they were obese. In follow-up tests five years later, the fatties scored even lower while the others scored as before. It is thought that those with a higher BMI had too much fat in their brain cells or that the arteries in their brains had thickened. So too much fat, especially the unhealthy saturated, hydrogenated and trans-fat varieties, might clog up the brain.

## Forget (with) these foods

Certain foods, drinks and substances are bad for your brain. Stay clear of fizzy drinks and sugary snacks, which produce an energy-sapping sugar low. Cakes, pastries and biscuits and other junk foods are high in trans-fats that clog up the brain (which is the organ with the highest concentration of fat at around 60 per cent).

Watch out for depressants such as alcohol, which severely affects our ability to form memories, and stimulants such as caffeine, which cause jitteriness and poor concentration. Many studies have also shown that consistent use of 'soft' drugs (such as marijuana) slows the brain because they contain fat-soluble substances that linger in the brain for weeks.

# GETTING YOUR BRAIN IN TRIM

The aphorism 'use it or lose it' definitely applies to the brain. It seems that if you don't use your marbles, you lose them. Get your brain in trim by stimulating and refreshing it so that it is open to new ideas and associations. Some studies have suggested that people who engage in regular cognitive activities like those below can reduce their risk of getting Alzheimer's disease. Some ways to do this on your own are by:

- ✔ Reading regularly.
- ✔ Doing crosswords (an excellent test of long- and short-term memory).
- ✔ Doing puzzles – both in books and those using manual manipulation.
- ✔ Doing Su Doku puzzles (particularly good for short-term memory and mental agility).
- ✔ Doing jigsaws, especially if you want to build your visual skills (see page 55).
- ✔ Learning to send (or speeding up at) text messages.

Activities that stimulate the mind and involve being with other people add a valuable social element.

These include:
- ✔ Visiting museums, art galleries and other stimulating environments.
- ✔ Learning new communication skills, such as using a foreign language, sign language or Braille.
- ✔ Learning moving skills such as ballroom dancing, drawing or playing an instrument.
- ✔ Joining a discussion group, such as a reading group. Talking about a subject helps you understand and remember it.
- ✔ Playing strategy games, such as chess.
- ✔ Playing card games, especially those in which you need to recall which cards have appeared, such as bridge and poker.

## Keep it up

Researchers in Germany found that people who taught themselves to juggle produced more nerve cells in the parts of the brain that deal with visual motion information. When they stopped practising, production of these cells slowed. So practising juggling – and presumably other coordination and movement skills – can boost your brain power, which will improve your memory.

Our brains are good at adapting to change. Change keeps us alert as to what is going to happen next, which in turn helps us to remember what happened before. Some ways to change your routine are:

- ✔ Try to use your 'wrong' hand for everyday activities, such as dressing, clicking the computer mouse, etc.
- ✔ Do the household chores in a different order.
- ✔ Load the dishwasher a different way.
- ✔ Eat a meal with chopsticks.
- ✔ Have an indoor picnic.

Seeing things from a new perspective activates the brain and the memory function. Try walking the 'wrong way' round the supermarket, or turning familiar pictures and calendars (even the clock!) upside down to alter your viewpoint.

### All change

Take a different route for a regular journey, such as the commute to work or the trek round the supermarket. You'll notice some things for the first time and even the familiar can take on a new look as you see it from a fresh viewpoint. This should stimulate your brain to examine input more carefully and to notice more.

# Memory and code

Working at creating or breaking codes and ciphers provides plenty of mental stimulation and can be great fun. It also requires excellent long- and short-term memory skills and an ability to manipulate stored information, for example using the alphabet in different ways. This is gold dust for memory skills.

## Sdrawkcab

Solve this backwards cipher:
**GNIDAER/A/EGASSEM/SDRAWKCAB/ SERIUQER/DOOG/YROMEM/DNA/ NOITARTNECNOC.**
Did you have to write it out or could you read it? Get someone to write out other backwards messages.

## Caesar shift

Try the Caesar shift, in which letters are replaced by those further on in the alphabet. Decipher this message, in which letters have been shifted along three places:
**WKH DQFLHQW URPDQV XVHG WKLV FRGH.**
Make up your own Caesar shift cipher message. Can you do it in your head without writing out the alphabet?

### Igpay Atinlay

Test yourself on a spoken code such as Pig Latin. The rules are:

1. Words starting with vowels have 'ay' added at the end.
2. Consonants at the start of words and double consonants are moved to the end and 'ay' is added.
3. Get someone to read this message to you and translate it:

   **'Igpay Atinlay akesmay ouyay ocesspray okenspay informationay icklyquay'.**

Make up your own Pig Latin message.

## Cognitive calculations

Some memory experts believe that performing a number of simple mathematical calculations every day tones the brain. These short 'mathematical sprints' are the mental equivalent of taking your body for a daily jog to build fitness.

## Some sums

Make up 30 simple sums using addition and subtraction, working with numbers up to 20, and multiplying any pair of numbers below ten. Answer the lot, making a note of your time. Try to beat your record by repeating the whole exercise (making up new sums) every day for a week.

## Counting race

Count aloud forwards to 120 as fast as you can. Now count it backwards. **120....**

| 119 | 118 | 117 | 116 | 115 | 114 | 113 | 112 | 111 | 110 |
| --- | --- | --- | --- | --- | --- | --- | --- | --- | --- |
| 109 | 108 | 107 | 106 | 105 | 104 | 103 | 102 | 101 | 100 |
| 99 | 98 | 97 | 96 | 95 | 94 | 93 | 92 | 91 | 90 |
| 89 | 88 | 87 | 86 | 85 | 84 | 83 | 82 | 81 | 80 |
| 79 | 78 | 77 | 76 | 75 | 74 | 73 | 72 | 71 | 70 |
| 69 | 68 | 67 | 66 | 65 | 64 | 63 | 62 | 61 | 60 |
| 59 | 58 | 57 | 56 | 55 | 54 | 53 | 52 | 51 | 50 |
| 49 | 48 | 47 | 46 | 45 | 44 | 43 | 42 | 41 | 40 |
| 39 | 38 | 37 | 36 | 35 | 34 | 33 | 32 | 31 | 30 |
| 29 | 28 | 27 | 26 | 25 | 24 | 23 | 22 | 21 | 20 |
| 19 | 18 | 17 | 16 | 15 | 14 | 13 | 12 | 11 | 10 |
| 9 | 8 | 7 | 6 | 5 | 4 | 3 | 2 | 1 | 0 |

## Counting jumps

Count aloud up in threes and fours from zero, and down from 500. See how far you can get in two minutes.

## Memory games

Lots of games require a good memory. As already mentioned, card games such as poker or bridge call on you to remember if certain key cards have been played. Here are a couple of non-card games that are particularly good for people of all ages.

### Fizz-buzz

Play Fizz-buzz on your own or in a group. Count in ones, saying 'Fizz' instead of every multiple of three (such as 6, 9, 12 etc.) and 'buzz' for multiples of five. For multiples of both such as 15 and 20 you have to say 'Fizz-buzz'. Regularly change the numbers for which you substitute words on multiples.

## Kim's game

Play Kim's game. Put 20 everyday items on a tray. Players have a minute to study them before the tray is removed or covered.
Now write down as many objects as you can in one minute.

## Picnic game

Play the picnic game. You need a few players for this. Each player takes it in turns to say 'I went on a picnic and I took…' repeating the objects the other players listed, in order, before adding a new item each time. You can make the game easier by having each item start with the next letter of the alphabet (apple, biscuit, coat, etc).

## Pelmanism

The memory game, also known as Pelmanism or Pexeso, requires a set of cards in pairs to be matched. You can use half a pack of playing cards by removing two of the four suits so you have two aces, two kings, etc. Lay all the cards face down in rows and columns. Turn over one card and study it and its position. Now turn over another and do the same. Turn both face down and repeat the process, trying to find the pairs and removing them as you do so. As you get better at it, increase the number of cards in play, and you can lay them out randomly rather than in rows to really test your skills.

## Pillow talk

Sleep is essential for our memory. When we sleep, the brain does not shut down. It goes over the events of the day, reinforcing memories and analysing problems. People learning a new skill can find their abilities improve overnight, because the brain continues rehearsing the new skill. The lack of new stimuli also offers some rest time so that our bodies (and brains) are refreshed. Conversely, lack of sleep inhibits brain power and reduces memory capacity. A nap while studying can improve your performance in an exam.

## SELF-AWARENESS

The better you know yourself, the easier it is to develop your skills, because you understand what you can and can't do, and appreciate what you'd like to learn to do and why. This allows you to develop an idea of the best route for you to achieve a better memory performance.

We all take in information in our own unique way. Teachers know that auditory learners have to speak or hear information, while visual learners need to see and/or write it. Others are kinaesthetic learners: they need to touch or feel something to understand it better, and for them it is easiest to find things out in a very practical way (like people who take equipment apart rather than read the manual). Most people tend to use a combination of sound, sight and touch (especially the first two), but it is always interesting to be aware of which we prefer, as that is likely to be the main route to a more reliable memory.

The exercises on the following pages will help you discover your learning preference(s). This self-awareness will help you boost your memory skills, as you can steer yourself towards the kinds of learning methods and associations (an information-linking technique, see pages 119–40) you feel most comfortable with.

## Window test

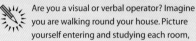

Are you a visual or verbal operator? Imagine you are walking round your house. Picture yourself entering and studying each room, counting the windows in each location. Which was easier, the imagined movement or the counting? The first element is carried out by the visualising part of the brain, the second by the verbalising part as you totalled up the numbers. Think about your own thought processes, and how you ponder problems. Visual learners tend to work more comfortably in pictures and diagrams while verbal learners prefer words. So two people can have totally different ways of remembering the same thing, and both can be effective.

## Auditory memory

This is the ability to remember things you have heard. Some people with good auditory memory only have to hear a joke once and they can repeat it word for word. Others will recall everything but the punchline, or the set-up, and find everything else has got mixed up or disappeared altogether. Some musicians only have to listen to a piece of music once, and they can play it, while others would struggle ever to achieve this without seeing notes on a stave.

## Pardon?

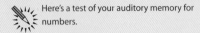
Here's a test of your auditory memory for numbers.

**Numbers**
2 21 13
3 15 22 54
4 45 64 7 18
5 99 53 72 67 4
6 76 50 23 1 62 68
7 64 92 16 8 72 39 77
8 60 73 62 17 19 43 88 35
9 44 81 20 37 64 89 57 73 63
10 82 47 28 58 93 85 60 22 7 26
11 36 23 15 4 96 72 55 76 46 12 90

Get someone to read you one row at a time of the random numbers from this table (keep it hidden from you). You must repeat the numbers, in order, before they read out the next row. Stop when you make a mistake. Seven numbers is an average score. Perhaps you are more a words than a numbers person. Will you remember a list of spoken words better?

The Pardon? exercise tests your short-term auditory memory. The next exercises test your long-term auditory memory. The first is a popular 'quiz night' activity.

## Name that tune

Set your ipod or other music playing system to play random songs. How quickly can you indentify the song's title? Can you say the first phrase of the song before you hear it? How does the chorus go?

This exercise tests your memory for specific voices and whether you can identify the subject of a conversation that has already begun.

## Who's talking?

Switch on a speech station, preferably a news programme, on the radio or television (so long as you don't look at the screen). If the subject or an interviewee is being introduced, turn off and wait before switching back on so that you are listening to a conversation that has already started. Can you identify who is talking? Do you recognise their voice or can you work it out from what they are saying? Can you work out what they are speaking about?

## Hear here

**Number of words**
- 2  food igloo
- 3  kitchen key pearl
- 4  insect pony goal phone
- 5  statue rack ship camera frame
- 6  face kiosk ball form zebra wheel
- 7  bugle sail text frame path oil grill
- 8  dial cargo foot prize castle month letter cloth
- 9  step test collar fish tree gate raft egg mine
- 10 mat cough dump fire spray jelly body court rag plank
- 11 canal toast plane kit flower jet stone satchel iron plug nut

Get someone to read you one row at a time of the random words from this table (hidden from you). You must repeat the words, in order, before they read out the next row until you make a mistake. Again, seven items is an average score.

## Visual memory

Some artists can study a building and produce a sketch or even an architect-style drawing of it without looking at it again. There are students who pay little attention in lectures (if they attend at all) but can mug up on the subject with a swift perusal of the notes. This is visual memory.

## Number vision

| 25 | 67 | 24 | 49 |
| 19 | 88 | 46 | 86 |
| 12 | 15 | 56 | 71 |
| 65 | 53 | 94 | 33 |

Look at this set of numbers for one minute then cover it up. Write down as many of the numbers as you can.

## Seeing words

| wall | scissors | cat | onion |
| radio | newspaper | circle | vase |
| gift | salt | music | coin |
| tip | theatre | hedge | pound |

Look at this list of words for one minute then cover it up. Write down as many of the words as you can.

Did you score more or less than when you were hearing or seeing the words or numbers? A major difference in scores suggests a bias towards that learning style, or that type of information – some of us can remember the digits of telephone numbers but not the words on a shopping list. Your ability to store words or numbers visually or orally will influence what you learn and how you remember it.

## A picture is worth…

Visual images work in a different part of the brain to words and sounds, and can be a valuable memory tool. They can help us understand information. For example, many children find maths problems easier to understand when there is a visual element such as a drawing, graph or diagram.

## Photographic memory

Someone with a photographic memory would be able to remember any visual image, and some people claim to be able to do this. But rigorous testing has never proved anyone capable of this. However, some people have a very good eidetic memory. This is when an image lingers in their mind for a few minutes after seeing it, and this can be confused with photographic memory. The consensus is there is no such thing as photographic memory, but some people have developed superb memory strategies that relate to their field of expertise. For example, top chess players can memorise many different chess boards with pieces in position – but when presented with impossible scenarios where the pieces are in illogical places, this ability disappears. So they are good at seeing patterns and remembering comprehensible images, but not random, illogical information. That exercise uses both your long- and short-term memory. The next one focuses on the short-term memory only.

### Picture look

Study a picture that you are familiar with, perhaps a painting that hangs in your house, or a family photograph. Give it to someone else and describe it to them in as much detail as possible. They can ask questions to clarify how accurately you can remember the image.

### In the news

Repeat the exercise in 'Picture look' this time with a picture you have never seen before, maybe from a magazine or newspaper. In addition to accurately describing the image, can you remember the words in the caption or story that accompanined it?

## SKILLS CHECK

We achieve more when we feel confident, and one way to feel better about ourselves is to celebrate how far we have come already.

> ### I'm good!
>
> Write a list of all the things you are good at, covering skills related to your job, hobbies and social life. For example, can you drive, speak confidently to large groups, paint well, play an instrument, create visual aids, run great parties, etc? How did you learn these skills?
> How did you commit them to memory?
> What things require prompts or reminders to achieve?
> Have they got any common characteristics?

This activity should give you a boost as you realise how many things you are good at. It may also prompt you to consider why you would like to improve your memory. Two of the most common reasons are a desire to be able to remember names, and to be able to study better and perform well in exams. Not everybody wants to be a world memory champion.

## Working memory

We can store about seven items (plus or minus two) of information in our working, or short-term, memory. This is why people trained to give computer-aided presentations usually limit the number of key points to no more than seven. The brain reviews these items on its 'mental jotting pad' and decides what needs to be retained and what can be filtered out and forgotten. After all, if we remembered everything that happened to us, our brains would become congested with masses of useless, irrelevant information and we would probably go mad. If a thought is discarded from the short-term memory it can't be retrieved: you have to 're-load' by taking in the information again.

### Seven up

What's on your mind right now? Write a list of the first seven things off the top of your head. This might include sounds you can hear, how you are feeling, some long- or short-term anxieties, whether you are hungry, what you've got to do today or tomorrow, etc. It is very likely that these memories will be prompted by a variety of sources from several senses. Repeat this activity after a few hours and see what has stayed and what you left behind.

## Issues

If the same issues or concerns keep cropping up, these are obviously ongoing thoughts that are occupying your memory. However, it is quite likely that some of your 'top of the head' thoughts were about the sights, sounds and smells around you. This is perfectly natural because the brain is constantly examining these sensory stimuli. The following exercise tests your ability to process information in your working, or short-term memory.

### The five Ws

This activity is easier with a partner, but can be done alone if you have a way of checking your answers – perhaps by recording the material you listen to. Listen to or watch a news story. Can you remember its key elements?
These are:
- ✔ What happened?
- ✔ When did it take place?
- ✔ Who was involved?
- ✔ How did it happen?
- ✔ Why did it occur?

Can you tell a partner this brief summary of the story, complete with answers to each question?

# Multi-sensory memory

We are multi-sensory beings: we experience the world through sight, sound, smell, taste, touch and emotions. These senses and sensations in turn contribute to our bank of memories, but some of these memories are only triggered by the sense with which we originally experienced or linked them. This activity should illustrate this:

### Autobiographical triggers

You are going to try to find triggers for autobiographical memories. If possible find real examples for each of the following triggers, otherwise use your imagination. Concentrate on each source with your eyes closed, taking time to explore it and see if your mind can retrieve any memories linked to each one:
- ✔ The sound of the sea (or any other watery sound).
- ✔ Touching the bark of a tree (or other natural textures).
- ✔ The smell of a banana (or other strong aromas such as a spice, peppermint or freshly-mown grass).

- ✔ Tasting a strong flavour, such as ginger, yeast extract or syrup.
- ✔ Anger (or another strong emotion).
- ✔ London (or another location).

It is likely that at least one of these triggers prompted memories from your past, possibly deep in your childhood. You may be able to explore these further using the same trigger. You can also try other similar triggers from the senses and emotions and see where they take you.

Continuing this theme of investigating the links between the senses and memory, we are going to explore each strand in depth.

## Sight

Sight is a crucial sense for memory and we are particularly good at remembering people's faces (even if we can't put a name to them). See also Kim's game on page 40.

We store visual information for about a tenth of a second, which is why we can 'write' with a firework sparkler: the image of the sparks in the air is retained, creating the impression that we are seeing a line rather than a flickering flame. You can test this with the following exercise.

## Image storing

Put your hand with fingers splayed in front of your eyes. The image beyond your hand looks jumbled and doesn't quite 'fit'. Now wave your hand rapidly from side to side, still with the digits splayed. The image will be clear. This is because your brain stores the previous image for a fraction of a second, allowing two images to be blended to form a more solid, coherent picture.

## Unseen jigsaw

Try completing a simple jigsaw without referring back to the picture on the box. Study the picture for a few minutes, then turn it over and put the pieces together to form the image.

## Spot the difference

Spot the difference puzzles test your visual short-term memory. See how fast you can find the ten differences between the two pictures (overleaf). The answer is on page 186.

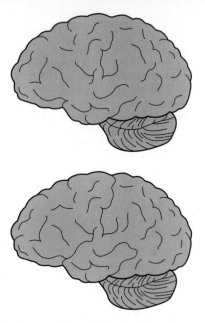

Now here are a few activities in which sight is removed from the equation. Instead you have to internally visualise your surroundings and what is happening, picturing where things are and being more aware of your other senses.

## Blind showering

Try blind showering. This is quite an easy way to start because we don't see all that much in a steamy shower anyway. Set the controls as normal so that you don't get the water too hot, and then wash yourself with your eyes shut. By closing your eyes you are making yourself more open to other sensations. You are also using your memory to recall where things such as the soap, shampoo and your towel are.

## Blind crossword

Try a blind crossword. It's OK, you don't have to memorise a whole crossword pattern! Get someone to read you crossword clues, providing the clue, number of letters and any that are known already. You'll have to picture how the word fits the available spaces. As the crossword builds up you'll have more letters already in place – and you may find that you'll come up with answers to previously heard questions as your brain whirrs through options without you being aware of it.

## Does this match?

Try blind dressing. Lay out your clothes then close your eyes and try to find each item and put it on. You are using visual recall combined with spatial awareness, coordination and your sense of touch.

## Where am I?

This is blind walking. Remove any dangerous obstacles that might trip you up. Then set a course to cross the room and pick an object up. Close your eyes and set off. Experiment with using just your visual image of the room and your position in it: try not to touch anything until you think you have reached your target. If this isn't possible, use touch as well.

## Chess problem

Set up a simple chess problem from a magazine or newspaper or choose one at **http://www.supreme-chess.com/chess-problems.html** or **http://www.bcps.knightsfield.co.uk**. Cover it and try to picture the board set up in your head. Now try to solve the problem.

## Sound

One of the keys to remembering things you hear is to eliminate any other sounds which distract your attention.

> ### Two-way listening
>
> Set up two different verbal sources, such as a television and a radio or audio tapes which you are not familiar with. Play them at the same time, and switch your attention from one to the other (close your eyes to avoid distracting images from the TV set). Listen for three minutes, shifting your focus between them, then note down as much as you can recall from both. If there is an imbalance, consider whether you are more receptive to different types of voice, or whether the subject matter of one was more interesting.

## Can't get you out of my head

Some tunes suddenly appear in your head for no apparent reason, and you can't get rid of them. These are known as earworms. Apparently we are more likely to remember songs with lyrics than music that is just melody, suggesting we remember more when we are looking for meaning. The songwriters of *Tin Pan Alley* used to listen to what the office cleaners were humming to find out which of their tunes were memorable. A survey of the most common recent earworms produced this catchy top ten:

1. Kylie Minogue: *Can't Get You Out of My Head*
2. James Blunt: *You're Beautiful*
3. Baha Men: *Who Let the Dogs Out?*
4. Theme to the film *Mission Impossible*
5. Village People: *YMCA*
6. Theme to TV programme *Happy Days*
7. Corinne Bailey Rae: *Put Your Records On*
8. Suzanne Vega: *Tom's Diner*
9. Tight Fit: *The Lion Sleeps Tonight*
10. Tiffany: *I Think We're Alone Now*

## Smell

Smell is the most evocative of the senses and enables us to trigger memories over a period of many years.

Human beings can tell the difference between up to 10,000 odours. We do this through special cells known as Blanes cells which, unusually, can themselves retain a memory, increasing the ability of the brain to recognise and make associations with the smell. This may explain why smells can so often trigger memories. Some scientists think that people who are not good at identifying smells are more likely to develop degenerative brain disorders such as Alzheimer's.

## Smelly words

Use a jar or a piece of cloth to store a distinctive smell, such as vanilla, peppermint or lavender. Create a random list of 15 words or objects and study them while breathing in the aroma. Try writing out the list, but don't check it. Wait half an hour then try again while breathing in the aroma. Now check both lists to see if smell helped you to learn the information. If it did, you created an association to aid your memory – see pages 78–84 for more on this.

## Aroma journey

You'll need a jar or cloth with a different aroma to the previous activity. On a journey, 'tag' ten places by smelling the aroma as you pass them, while noting down the location (it is better if you get someone else to do the writing). At your destination, take a sniff of the aroma and see if you can recall all ten points on the journey.

## Touch

Here are a couple of exercises illustrating that we can be very good at remembering what things feel like. The second task, identifying coins, is trickier.

### Touch sensitive

Make (or better still, get someone else to collate) a collection of at least ten different small items, such as a paper clip, nail, screw, pair of manicure scissors, button, coin, bead, battery, etc. Put them in a bag and identify them by touch. You can try this activity while distracting yourself with another activity such as talking on the phone.

### Money problem

Collect a selection of coins (more than one of each type) and see if you can identify them by touch in a bag or pocket.

We combine touch with spatial awareness when tapping out phone numbers and text messages, typing on a keyboard or playing a keyboard musical instrument. The skills are very similar: many pianists find they pick up computer keyboard skills very quickly as they already have highly developed hand/eye coordination and spatial memory.

## Tap safe

Study the keypad of the mobile phone for two minutes. Try tapping out your name while keeping your eyes fixed on the screen or, if you are feeling brave and confident, with your eyes shut.

## Finger clicking good

Type the alphabet on a computer keyboard without looking at your fingers. Repeat with your eyes shut. Once you are successful, time how fast you can do it without mistakes.

### The escalator phenomenon

We sometimes see what we expect to see, and feel what we expect to feel. Most of us have experienced that odd sensation when we step onto a stationary escalator and lurch forward as if we are still expecting the apparatus to move. Because you remember how the escalator should feel, your brain sends out signals telling you what to expect – then you get a surprise because what you expected doesn't actually happen.

## Taste

It is possible to train your palate to detect and remember different flavours (think of a professional wine writer). Taste is closely related to smell and can trigger memories from long ago as old circuits are re-activated. In childhood we experience many tastes for the first time and develop strong feelings about certain foods – either about the sensation we feel in our mouths tasting them, or emotional associations such as not enjoying a meal served by someone we didn't like. This may explain why we become attached to eating particular meals at the same time every year: we could eat turkey on any day, but many of us only eat it at Christmas.

### Make a meal of it

Think of the type of meals or sweets you ate in your childhood or at a major event such as a wedding, special anniversary or on holiday. Prepare the meal (or buy the sweets, if they still exist) and eat slowly, letting your mind wander to discover if old memories are ignited.

## Emotions

Emotional memory comes in many forms. We might remember the intense feeling we had at crucial points in our lives: first crushes, love, fear, depression, joy, etc. It might be linked to certain places, such as the location of a car crash, or a church or hospital, where something momentous happened. Emotional memory can also be triggered by the other senses.

### Flashbulb memory

 Choose several events from the list below. What were you doing when you heard the news? Can you remember what you were looking at? Who were you with? How did you feel? How did others react? Many people can remember where they were when they heard shocking news such as:

- ✔ The death of Princess Diana.
- ✔ The death of Elvis Presley.
- ✔ The shooting of John Lennon.
- ✔ The assassination of John F Kennedy.
- ✔ The 9/11 New York atrocity.
- ✔ The 7/7 London bombings.
- ✔ The Challenger space shuttle crash.

Often these memories include how you felt at the time and how others reacted. The intense feelings associated with hearing about such events leave a strong impression on our brain: we appear to remember them especially vividly. The memories may also be reinforced by the regular media references to these events, which rekindle our own recollections.

# MEMORY TRIGGERS: STARTING TO REMEMBER BETTER

There are a number of practical steps you can take to improve your memory without learning any special techniques. You can change the way you take in information that you know you will want to recall, and introduce habits that will help you to trigger your memories.

## Remembering names

First, what many of us wish we could do better: remembering names and putting names to faces. This is an important social and business skill, because people are more likely to want to engage with you if you call them by their name. There are benefits in forming friendships, building relationships and making deals.

Some names are easier to remember than others. These are the more unusual ones, which tend to trigger an immediate association so that we don't have to work as hard to install the information. For example, Melody suggests music, Daisy is also a flower and Brooklyn is a place. Some people light up the room and we tend to remember their names because of their charismatic personality. Of course, the opposite is also true: some people don't make an immediate impact because they are shy or overshadowed by others. Teachers have 'invisible' children in their classes who deliberately avoid attracting attention. This exercise will identify the 'invisible' people in your life.

> ### Spot the invisibles
>
> Think of any group of more than ten people you know. They could be fellow students, members of a team or social group. Write their names out. Check the list. The people you missed out, or struggled to recall and wrote down last, are the 'invisibles'. This doesn't make them less interesting, but they are likely to be the self-effacing personalities who are more comfortable one-to-one than in large groups.

When you are going to meet new people it is a great advantage to be familiar with their names in advance, because then your memory has something to build on. You can do this by asking your host in advance who else is coming or by reading through a list of fellow-delegates at a conference. If you are able to pick up other information in advance (such as where they live, who they work for or who they are married to) so much the better: it means you are better equipped for the actual introduction.

## Meet, greet and repeat

Sometimes we don't remember names because we weren't really trying: a distraction elsewhere in the room caught our attention at the crucial moment, or we were thinking about the last or the next person to be introduced, or something else altogether. It pays to make a real effort first time round. This avoids the common occurrence of learning the wrong name for someone: the mistake is likely to be repeated several times before you manage to 'unlearn' it. Take your time: this way you give yourself a genuine opportunity to input the name before moving on. Use all your senses (looking, listening, smelling perfume or aftershave, remembering the feel of the handshake).

- ✔ Listen to the name. Depending on the setting, you'll know whether you need to focus on the first name or surname.
- ✔ If you didn't hear it properly, ask now – much more socially acceptable than forgetting it (suggesting a lack of interest) and having to ask for it later.
- ✔ Say their name back to them ('Nice to meet you, Tim'). If possible, use it again during your conversation, and certainly when you part.
- ✔ If appropriate, ask for help in pronouncing or spelling the name – both could provide a valuable memory trigger.
- ✔ Look at the person's face, especially the eyes.

Are any of their features memorable (colour of eyes, type of complexion or the shape of the ears or nose)? Beware of relying on hair style or colour as it can change. Similarly, spectacle-wearers might only wear glasses on certain occasions, or could switch to contact lenses.
- ✔ Can you make any immediate link between them and their name? Maybe you could picture its first letter written on their face, or make some other link to do with appearance ('Bob: bushy eyebrows') or personality ('excitable Eve').
- ✔ If you swap business cards, make notes on the card itself.

### File it

Some people keep card files (or electronic versions) on which they record details of contacts. Making a note of personal information, such as a birthday or size of family, can be handy as you can refer to such details in your dealings with that person. This can be a valuable sales technique: we buy more from those with whom we feel an affinity.

## Remembering everyday things

Routine can be extremely valuable in helping your memory in two ways. First, if you often can't find your car keys or your phone, get into the habit of always putting them in the same place – keys by the front door, phone by the television, kid's homework in the hall, etc. This might seem very obvious, but some people need to train themselves to do it. Habits evolve, too: if you have to start taking regular medication at certain times, keep the pills by the kettle or the toothpaste so that you see them at the right point in the day. Sometimes the time of day or a radio programme can be a useful trigger: doing your exercises while the evening news is on, or taking your pills when you hear the theme tune to a favourite programme. If there is no convenient prompt that is part of your routine, set your oven timer or mobile phone and use it as an alarm clock (or just use an alarm clock!).

The second benefit of habits such as this is that you can play with the routine to create a trigger: if you've got a crucial journey to make next morning that is not in your normal routine, put your car keys in the fridge next to the milk so that you notice them as you have breakfast. A less dramatic variation is to put the alarm clock out of reach so that you can't switch it off and go back to sleep 'on automatic', or leave your phone in your shoes to remind you to make a call in the morning. Placing objects in unfamiliar places is a valuable memory technique: we look at it later in the book as a mental strategy.

A classic variation of this is to tie a piece of string or ribbon to your finger to act as a trigger for the memory. This isn't always practical or desirable, but you could try these other possibilities:

✔ Wear your watch on the opposite wrist or turn it so it faces away from you.
✔ Write a note on sticky paper and attach it to the fridge or the car dashboard.
✔ Put your keys in a different pocket.
✔ Twist your ring so that the stone is on the underside of your hand.
✔ Attach a 'sticky' to your wallet or purse.

## Notes and lists

Leaving reminder notes is also effective, especially in places such as a kitchen noticeboard, or on the fridge. The most obvious place for noting appointments and other events is in a diary or on a calendar, provided you make a habit of checking it. There are also electronic organisers, computer software and websites such as **www.eventprompt.co.uk** or **www.etimeinc.com** where you can log important reminders so that they will transmit reminder text messages or emails.

Another strategy some people rely on is list making. Sometimes the act of writing the list serves as sufficient reminder and there is no need to refer to it again. Write shopping lists in the order in which you will visit those shops or aisles, so that you encounter the items in the same order. If you tend to rely on lists to remind you what to do, keep a notepad on hand all the time, so that you can scribble down other reminders as they occur to you, rather than wait until the end of the day or the next morning to write your 'to do' list.

Some people make notes on personal recorders, or phone themselves up and leave a message – this is a good one for those sudden thoughts that occur to you while driving the car, provided you have a hands-free phone. At other times, you could send yourself an email or a phone text.

If you have difficulty remembering appointments or getting organised ahead of scheduled events, pack your bag the night before and leave it by the door: it will act as a reminder of what you will be doing, and you won't have a last-minute panic to get ready for it.

### Untidy desk, untidy mind?

Some people's desks always seem to be immaculate, while others resemble a volcano of paperwork piled up. But are the 'filers' better organised and less liable to forget things than the 'pilers'? Not necessarily, apparently: the piler's mountain of papers can form a set of cues that lead them to what they need – so can actually be a memory aid. Tidy up and they're lost!

# What if you still forgot?

It's a terrible feeling when you realise you've forgotten some crucial thing – such as where on earth you put the car keys. One way to address this is to mentally retrace your steps, recalling where you were, what you were carrying, what you did next and so on until you remember placing the umbrella, keys, phone or whatever. Retracing is a skill that can improve your general memory performance. For example, if you've lost your wallet or purse, you would work from the last time you were in a shop, which is the last time you are likely to have used it. Try this exercise:

> ### Retracing
>
> 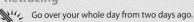 Go over your whole day from two days ago. Try to remember anything you bought (Where? How much did you pay?) and what you ate (What? Who with?). Recall three events in a chronological sequence after each of these.

### Action on distraction

It may be that distractions – such as TV or radio programmes, music playing, habitual texting or chatting with colleagues or on the phone, frequently checking for emails – could be having an effect on your memory. There could be environmental factors such as an uncomfortable temperature or noise from outside. These could all affect your capacity for taking in or recalling information, especially if you are not a particularly well-organised person. Try removing distractions, and set yourself a time or a target for concentrating. Resist the urge to check frequently for emails and texts – do it every hour or two. Ridding your life of distractions could free up your memory.

### Take a blind bit of notice

When we are focused on one task, we can miss significant changes around us. For example, in one study, a researcher asked a stranger in the street for directions. While they were concentrating on giving the answer, accomplices would carry a door between the pair, and behind this temporary block another person would replace the original questioner. Half of the subjects did not notice the switch – they were concentrating too hard on their task to notice the change. Similar results were recorded in a variation of the experiment when hotel guests failed to notice that the receptionist dealing with them had been replaced during a distraction.

# WAKING YOUR BRAIN UP

## Introduction

We remember what is memorable. Some things are immediately memorable: for example, most of us can identify Italy or Australia on a map or globe because they present a recognisable shape of a boot and a face respectively. We find this distinctive, which makes it memorable. What shape is Belgium, Thailand or Chile? Unless you had a reason to know, you probably don't have a clue.

Similarly, most of us might recall that Sir Walter Raleigh allegedly covered a puddle with his cloak to prevent Queen Elizabeth I from muddying her shoes, or that he introduced England to the potato and tobacco. These are quirky 'facts'. Most of us wouldn't know that he was a major Elizabethan poet, or that his explorations had an important part in the growing colonisation of America. These facts don't harbour the same human interest and they are not as memorable unless you are a historian.

The key to building your memory skills is to make the information you want to retain memorable. In this section we start to look at how to find patterns and create associations, and in particular how to use mnemonics. These skills are particularly useful for studying, and the section finishes with advice on how to learn and retain information efficiently.

### Why the rote route is a dead-end

Traditional schooling used the rote learning method of constant repetition to sear facts into children's brains. Classes would chant multiplication tables or the alphabet until they were word-perfect and appeared to know these 'facts'. However, the problem with rote learning is that it does not foster understanding, so children often can't use the information effectively. This is why some children, asked which letter is three after 'T', have to recite the alphabet from the start, or why a child can chant a times table flawlessly but cannot answer questions using its content. There is no meaning attached to the learning: there are no associations that enable children to understand and use the facts they 'sing'.

# MAKE IT MEANINGFUL

We learn and remember most either when we find the process enjoyable (good tutors make learning fun) or are stimulated by the information (needing it to pursue a career or a hobby, for example). The more you know about a subject, the more you can learn, because you have an understanding of what has gone before, and this breeds links and associations as you find out more. So the information is meaningful, and parts of your brain are already storing similar data, which you supplement. This exercise illustrates that learning must be meaningful.

> ### Nice list, nasty list
>
> Write a random list of ten unrelated words from the dictionary, choosing a new page for each word and selecting it by chance (eyes closed, place your finger on the page and find the nearest word). Now write a list of ten words to do with your job or hobby. Study both lists. Cover them and try to write them out. You are most likely to best reproduce the list relating to your own interest because it has meaning for you – even though it might not to someone else.

# FIND PATTERNS AND ASSOCIATIONS

One way to make information more meaningful is to find patterns or make associations in it.

## Patterns

Look at the number sequence: **3810151722242931363843**. It is very hard to memorise this until you recognise the pattern 'add five, and then add two'. All you need then is the starting number (three) and you can re-create the sequence. Obviously, a random sequence is unlikely to have such a pattern but this is a good method for creating memorable number passwords.

The method works for words, and therefore for letter passwords too. Try memorising: **JAANILENPTHITOTCBUETWAR**. Since it makes no sense, it is extremely difficult to learn because it feels pointless. Once you know it is created from alternate paired letters of the nursery rhyme **JACKANDJILLWENTUPTHEHILLTOFETCHABUCKET OFWATER** it becomes readily memorable – you can trigger the sequence just by thinking 'Jack'.

## Tell me on a Sunday

You can use the patterns relating to a number to perform an impressive party trick. Start by memorising the sequence **265274263153** (you may find it easier to learn it as the four three-digit numbers **265**, **274**, **263**, **153** – see 'chunking' on pages 104–12). From this sequence you can tell people on what day events happened in the year 2000, such as the day their birthday was, or whether Christmas Day fell on a weekend. It works like this: each digit is the date of the first Sunday in the month. The first digit is two, so 2 January 2000 was a Sunday, the next digit (six) means that the first Sunday in the next month was 6 February, and so on. Armed with this information it is simple to count on to find what day of the week any date was. The first Sunday in December was the 3rd, so the Sunday before Christmas Day was the 24th (3rd-10th-17th-24th), and Christmas Day was the next day, a Monday. If you have the time and inclination, you can create a 12-digit sequence for any year and widen the scope of your trick.

> ### Phrase trigger
> 
> Find a phrase that is useful in a subject you are studying or relating to your job (or, for the purpose of this exercise, just choose a nursery rhyme). Create a sequence of letters in the same way as the example above, then practise writing the sequence out working from a single word 'trigger'.

Another way of finding patterns from numbers is to use the format of keypads such as those found on telephones or devices for entering your PIN number in shops and at bank machines. These usually have three columns with numbers appearing in sequence as you read across, the '0' being on the bottom row alongside the 'cancel' and 'enter' buttons. Four-digit PIN numbers are easy to learn, and we get so used to entering them on a keypad that we can often do it 'blind', relying on our spatial awareness to hit the right key. So the PIN number has a shape: the movements your fingers make to enter it. A shape that works round the corners would key in 1397. The line going straight down the centre column forms 2580.

# ASSOCIATIONS

Another way to create meanings for numbers is to find a popular association with them. For example, the sequence **74736052007365** would be very hard to remember, but it could trigger these associations:

- **747** is the number on a Jumbo jet.
- **360** is the Xbox 360 console game
- **52** is the number of cards in a standard deck
- **007** is spy James Bond's number
- **365** is the number of days in a non-leap year.

So you could trigger the sequence with the words 'Jumbo, Xbox, Cards, Bond, Year'. You don't even necessarily have to remember these words: we remember visual information such as pictures better than words, so you could create a picture with all these elements, such as a jumbo jet with a calendar painted on it and James Bond sitting on top of it looking at a stack of cards on his Xbox.

This method can work well for words, too, creating a picture linked to the word and adding images for the other words in the list. This technique works much better when the words describe tangible things such as concrete nouns (things you can see or touch) that have clear visual meanings. Abstract nouns such as 'truth' or 'idea' are trickier because you have to work harder to create an image for them.

## Associated sequence

Think up other numbers with ready associations, such as Levi 501, Porsche 911, 45: the speed at which vinyl singles are played; 10 Downing Street or the famous book and film The 39 Steps. Make up a sequence and see if you can remember it from the phrase you create, or from a picture that contains trigger images.

## Mental pictures

Create a mental picture for each word to remember this list:

| | |
|---|---|
| tree | idea |
| wall | light |
| smile | belief |
| car | truth |
| book | generosity |
| newspaper | help |

It should be much easier to picture and recall the first six because they are concrete nouns, while the other words are more abstract and harder to visualise.

# False associations

The associating technique works because the brain is always ready to 'play' and make links with new information because this fosters understanding. Sometimes the brain gets carried away. We can illustrate this:

## Spot the fake

Study this list of words for two minutes. Then cover it and write down as many of the words as you can (don't worry about missing a few out).

**painting oil watercolour picture frame canvas landscape portrait colour sketch outline mix palette coat stroke**

Now repeat the exercise with this list:

**watch select channel programme serial screen remote satellite station guide review repeat volume series digital**

Do not read the following text until you have written as many of the words as you can.

> Check the two lists you wrote and see if the first one contains 'brush' and the second one includes 'television'. Neither of these words appears in the lists, but since it would be perfectly reasonable to expect them to as they fit the categories of 'art' and 'television', your brain is quite likely to have decided they should be there. Including the missing words is not a sign of a faulty memory, but of a brain keen to make associations. Some people who have taken part in this test have argued with the researchers that the lists have been changed and the 'phantom' words were really there.

### False memory syndrome

Remembering things that did not happen is called false memory syndrome and is a controversial topic in psychotherapy. There have been cases where adults have made serious allegations about abuse during their childhood that they seem to genuinely believe are 'recovered memories', but which turn out to be untrue. Of course there are many occasions where such memories are real, but it seems that it is possible for our brains to manufacture past events, perhaps as a result of being prompted by leading questions from others.

# BETTER SPELLING

Some people find it very hard to spell accurately. A common way of addressing this is to generalise a spelling rule. For example, people know 'i before e except after c' and can apply this rule rather than learning every word with this letter combination. Other strategies to try out that may help include:

- ✔ Find words within the word (there's a 'hen' in 'when'!).
- ✔ Break the word up into smaller parts (Wed + nes + day = Wednesday).
- ✔ Break the word up into sounds (comp-le-ment-ary).
- ✔ Write the word with your finger in the air, or in sand.
- ✔ Type the word, or pretend to.
- ✔ Say the word as it is written (like clim-b or k-nife).
- ✔ Trace over the letters with the finger several times, saying the sounds (combining kinaesthetic, auditory and verbal input).
- ✔ Find a word that rhymes with it: is the spelling the same?
- ✔ Count the letters – if you remember there are 13 letters in 'accommodation' you are more likely to include the double letters.

Another useful spelling-reminder method brings us to a key memory-improvement technique: mnemonics.

# SIMPLE MNEMONICS

A simple mnemonic is a word or short phrase that helps you remember something because it acts as a key to the lock that reveals a stash of information. This technique can be used in many ways, and mastering some of them will take you a long way on the road to improving your memory.

## Mnemonics and spelling

Some spelling mnemonics are:

- ✔ **Because**: 'Big Elephants Can't Always Use Small Exits' (this is an acrostic sentence, made using the letters of the word).
- ✔ **Diarrhoea**: 'Dash In A Rush Really Hurry Or Else Accident'.
- ✔ **Necessary**: 'Never Eat Cakes Eat Sardine Sandwiches And Remain Young' (or, concentrating on the 'c' and 's' issue, the simpler 'one coffee, two sugars').
- ✔ **Rhythm**: 'Rhythm Helps Your Two Hips Move'.
- ✔ To delineate **deserts** and **desserts**: the sweet one has two sugars.
- ✔ Finally, the tricky spelling of the word itself is covered by 'Monkey Nut Eating Means Old Nutshells In Carpet'.

The more surreal and bizarre the content of the mnemonic, the more likely we are to remember it, because it becomes fun and is unlikely to get confused with other information.

Mnemonics are used in many disciplines. Some examples are:

**From astronomy:**
to remember the order of planets in our solar system. **My Very Efficient Memory Just Speeds Up Naming Planets** or, now that Pluto has been relegated in status: **My Very Energetic Mother Just Screamed Utter Nonsense**.

**From geography:**
**Never Eat Shredded Wheat,** or **Naughty Elephants Squirt Water**, to learn the points of the compass in order.

**From geology:**
**Stalagmites might reach the roof, stalactites have to hang on tight or they will fall off.**

**From the sciences:**
for the spectrum (commonly used to describe the colours of the rainbow): **Richard Of York Gave Blood In Vain**. A more contemporary version is: **'Run Off Young Girls, Boys In View'**.

To summarise the seven life processes:
**Mrs Nerg**, which stands for Movement, Reproduction, Sensitivity, Nutrition, Excretion, Respiration, Growth.

In mathematics:
the order of operations is set by **Bodmas**: **Brackets Over Division Multiplication Addition Subtraction**. Another reminder is **My Dear Aunt Sally: Multiply and Divide before you Add and Subtract.**

To remember the digits of Pi:
just count the number of letters in each word: **'How I wish I could calculate pi'**, which neatly provides seven digits, or the longer: **'How I like a drink, alcoholic of course, after the heavy lectures involving quantum mechanics'.**

A (perhaps older, and certainly more refined) rhyme giving the same information is:

> **Pie**
> **I wish I could determine pi.**
> **Eureka! Cried the great inventor.**
> **Christmas pudding, Christmas pie**
> **Is the problem's very centre.**

**From music:**
**Every Good Boy Deserves Favour** (for the notes E, G, B, D, F resting on the lines of the treble clef).

To remember whether we lose or gain an hour adjusting from summer time: **Spring Forward, Fall Back** (using the American term 'fall' for autumn).

To remind us of the number of days in each month:

> **Thirty days has September,**
> **April, June and November.**
> **All the rest have thirty-one**
> **Excepting February alone,**
> **Which has twenty-eight days clear**
> **And twenty-nine in each leap year.**

## A kinaesthetic mnemonic

There is another mnemonic for remembering the number of days in each month. This is interesting because it uses the hands rather than words.

Clench both hands into fists with the knuckles facing you (so palms facing out) and notice the rise and fall contours of the knuckles separated by the hollows in between. These represent the months of the year. Starting on the left fist:

**January:** The knuckle of the left (fourth) little finger.
**February:** The hollow to the right.
**March:** The knuckle of the ring (third) finger.
**April:** The hollow to the right.
**May:** The knuckle of the middle (second) finger.
**June:** The hollow to the right.
**July:** The knuckle of the index (first) finger.
On the right fist:
**August:** The index knuckle.
**September:** The hollow to the right.
**October:** The middle finger knuckle.
**November:** The hollow to the right.
**December:** The ring finger knuckle.

You could just use the left fist, starting again on the little finger knuckle for August. Knuckle months have 31 days, while hollows are shorter (30 apart from February). This method is popular in many parts of the world including China, Russia and much of South America.

# Timeless mnemonic

A necessarily very long mnemonic for remembering the geological time scales is as follows:

| Word | Geological period |
|---|---|
| pregnant | Precambrian |
| camels | Cambrian |
| ordinarily | Ordovician |
| sit | Silurian |
| down | Devonian |
| carefully | Carboniferous |
| (most peculiarly) | (Mississippian Pennsylvanian) |
| perhaps | Permian |
| their | Triassic |
| joints | Jurassic |
| creak | Cretaceous |
| perhaps | Palaeocene |
| early | Eocene |
| oiling | Oligocene |
| might | Miocene |
| prevent | Pliocene |
| permanent | Pleistocene |
| hobbling or | Holocene or |
| rheumatism | Recent |

## Song and rhyme mnemonics

Songs and rhymes are valuable because they make the learning fun and the links offered by melodies (which some people remember very well) and rhymes (which trigger associations with other words) help us to remember.

Many children gain familiarity with multiplication tables through songs and learn the alphabet by singing it to the tune of 'Twinkle Twinkle Little Star' in the song that goes:

**AB, CD, EFG**
**HI, JK, LMNOP**
**QRST, U and V**
**WX and Y and Z**
**Now I know my ABC**
**Next time come along with me.**

### Remember us!

Organisations try to make themselves memorable by giving themselves titles whose initial letters spell real or imagined words (acronyms) – and become used as words themselves. Examples are National Aeronautics and Space Administration (NASA); United Nations International Children's Emergency Fund (UNICEF). Acrostics are everywhere, examples being Acquired Immune Deficiency Syndrome (AIDS); Radio Detection and Ranging (RADAR); Light

Amplification by Stimulated Emission of Radiation (LASER), Self-Contained Underwater Breathing Apparatus (SCUBA). Their widespread use highlights the value of reducing any piece of information to as short a form as possible.

## Creating simple mnemonics

A good way to start creating and using mnemonics is for postcodes, knowledge of which can sometimes come in very handy. Take the postcode **SW1A 1AA**. A possible mnemonic for it would be **Super Woman One, A One Alpha Adult**, which would be appropriate as it is the postcode for the royal residence Buckingham Palace in London. Any such mnemonic is likely to be more useful if you can link it to the person concerned.

### Royal mail

Make up and learn mnemonics for these postcodes: **SL4 1NJ** and **AB35 STB**. They are the codes for the royal residences at Windsor and Balmoral. Try to give your mnemonics a regal flavour.

Of course, the best way to learn to apply mnemonics is to devise and use your own for information you want to learn. That provides motivation and you can have fun making your mnemonic appropriate to its subject matter.

> ### Melodic memories
>
> Find any set of information that you are interested in and devise your own mnemonic for it. If it appeals, include rhymes or set it to a tune. Make the content as odd or appropriate as possible so that it is genuinely memorable. Can you reduce it to a one-word acronym?

The combination of words, melody and rhythm creates many associations for the brain to use and explains why we can remember songs and sung mnemonics (such as the children's ABC song and the 'Pie' rhyme) so well. This was and still is the case in many cultures around the world. For example, in many parts of Africa drums are used as mnemonic aids, and indeed variations in pitch communicate information in addition to the words being spoken.

## AS SIMPLE AS ABC

The alphabet is a valuable memory tool. We are used to using the alphabet to store and find information via phone books, dictionaries and the like. Remembering information alphabetically allows us to store it efficiently and access it quickly.

> ### Alphabet of nations
>
> Write an alphabet of nations: one country for each letter of the alphabet, in order (you can miss out X, and there is a full list at **http://dir.yahoo.com/Regional/countries/**). Learn it and get someone to test you. Start by reciting the list from the start. Then start from anywhere in the alphabet. Finally, get them to ask you for single countries in random order.

In this exercise there is one 'trigger' – the letter of the alphabet – for each word. The next step is to build on this by adding to the number of individual entries for each letter.

## Shopping ABC

There are 20 items on this shopping list: **chocolate, bananas, spaghetti, sugar, cheese, milk, mushrooms, flour, beans, bread, fruit tea, mango, coffee, fennel, coconut, flowers, salmon, sausages, butter, mustard.**

You are very unlikely to be able to remember the list. However, there are only five initial letters (**b, c, f, m** and **s**) each with four items. The re-ordered list reads:

**bananas, beans, bread, butter,
cheese, chocolate, coconut, coffee,
fennel, flour, flowers, fruit tea,
mango, milk, mushrooms, mustard,
salmon, sausages, spaghetti, sugar.**

Memorise this list and get someone to test you on remembering each item in each group (don't worry about the order).

You were helped by knowing that each group had the same number of entries. The following exercise removes this advantage, to continue to build your memory.

## Straight As

Study this list of twenty countries:

**Angola, Argentina, Australia, Bolivia, Brazil, Ghana, Germany, Greece, India, Iran, Italy, Norway, Pakistan, Singapore, South Africa, Spain, Sweden, Syria, Thailand, Tunisia**.

Notice and remember that there are three As, two Bs, three Gs, three Is, one N, one P, five Ss and two Ts. Test yourself on recalling these. If this is too challenging, simplify the list to, say, 12 entries and build up from there.

The next step is to try to remember groups of words for letters in the alphabet, and store them alphabetically by second or third letter.

## From Aaron to Axel

Think of the people you know whose name starts with A. 'File' them in alphabetical order: Angela, Andrea, Abbie, Alex, Adam, Aron, Andy would be re-ordered Abbie, Adam, Alex, Andrea, Andy, Angela, Aron. When you have about 20 names, test yourself on your recall in order. It may help to check how many names there are for each second letter, so this list would be 1 x b, 1 x d, 1 x l, 3 x n, 1 x r.

# CATEGORIES

So far, the alphabet has been serving as a way of grouping items into small sets of information which can be triggered by one letter of the alphabet. Now you can expand this idea of putting things into groups by using other categories. For example, the shopping list used in the alphabet exercise could be re-grouped as follows:

✔ **Dry goods:** beans, spaghetti, sugar, chocolate, bread, flour, mustard, coconut **(8)**
✔ **Fresh fruit, vegetables and flowers:** bananas, fennel, mushrooms, mango, flowers **(5)**
✔ **Dairy:** butter, cheese, milk **(3)**
✔ **Drinks:** coffee, fruit tea **(2)**
✔ **Meat/fish:** salmon, sausages **(2)**

You can play with the categories, perhaps putting the dairy with the meat and fish items by grouping them as 'from the chiller', or adding the drinks to the dry goods. Putting the groups into the order in which they are found at your supermarket or on your high street also makes the list more effective in practice. The point is that rather than trying to immediately recall every item on the list, you have now created a small set of triggers such as Dry 8, Fresh 5 and so on.

## Continental shelf

Read the list of countries in the alphabet exercise on page 97, taking your time and thinking about the continent each one is on as you read. This information is summarised in the table below:

| Continent | Countries |
|---|---|
| Africa | Angola, Ghana, South Africa, Tunisia **(4)** |
| America | Argentina, Bolivia, Brazil **(3)** |
| Asia | India, Iran, Pakistan, Singapore, Syria, Thailand **(6)** |
| Australia | Australia **(1)** |
| Europe | Germany, Greece, Italy, Norway, Spain, Sweden **(6)** |

Try to learn this information using the continent and the number of countries listed for it as triggers.

This technique of reading and immediately categorising is called 'active reading'. Because you are making judgments as you read, you are processing the information more carefully and thoroughly, and (with practice) 'filing' it efficiently. This is called 'depth of encoding'.

We use categories naturally in many areas of our lives, for example differentiating people between work colleagues, social friends, team mates and so on – already there is likely to be some overlap at times.

## Categorising people

If you wish to remember the names of a large number of people, you could try to set up a sort of mental filing system where you can put new acquaintances into the 'right' file as you meet them. For example, you might put all the Mikes in one folder, all the Megans in another. Alternatively, especially for business acquaintances, you might file them by their occupation or role: suppliers, clients, or (if these are too broad) printers, packers, distributors, and so on.

> Flick through your contacts book and mentally file people by first name or by their role in your life. You may choose to annotate your book with these categories (using colours or initials). Test yourself on your recall. If you are using the first name method, list all the 'Mikes' and 'Megans' in separate files.

## Multi-sensory grouping

Bearing in mind that memory is multi-sensory, you could employ different senses to recall items in groups. The next activity allows you to experiment with this idea by deliberately linking words with their related senses.

### Make sense

Study this list of 16 words and try to remember them using the senses suggested, so that you learn four groups of four, each linked by one of the senses.

| Sense | Things to remember |
|---|---|
| Sound | Laughter  Music  Crying  Running water |
| Taste | Vinegar  Mustard  Chocolate  Salt |
| Smell | Peppermint  Washing powder  Coffee  Banana |
| Touch | Wood  Keyboard  Glass  Pillow |

This should help you to identify if any of these senses are particularly valuable for you as routes to memory triggers. If you found it really easy to conjure up one of the groups and were reminded of other suitable words or ideas for them, you might consider using this sense to help you form links and associations (see pages 113–40).

## Studying in colour

One way of learning a set of study notes is to decide the different categories information falls into and allocate a colour to each, then mark up the text with crayons or felt-tip pens. So if you decide everything historical will be red, and anything relating to future trends is green you will be able to identify the relevant data quickly. This might form the basis for a fresh, re-ordered set of notes, or you might find you can scan the text, picking out what you want from the colour markings.

### Colour connections

As has been shown, our brains remember things using association, so it can also help to draw coloured lines across notes, linking topics or ideas together. You could choose a separate colour for these lines and visualise that colour when you are 'searching' for information in your mind. Alternatively, draw double lines using both colours, or combine them (for example, red and yellow categories could be linked by orange).

If you make or keep most of your notes on a computer rather than using pens and paper you can still use this method by changing the font colour as appropriate. This makes it very easy to skim through notes when seeking out specific topics. You can then use these colours as triggers to recall the information for each category. As you come across new learning, you can mentally file it in the right colour folder.

> ### Colourful notes
>
>  Find some written text about a topic you are interested in or need for work. Identify three to five categories for the information and colour code it. Read the text using each colour in turn, making notes if you wish. Concentrate on linking the category with its colour. Revise your notes after a day and again after a week. Then test yourself by writing out the colour-coded information by category. Check how much you recalled against the original.

### A coloured ABC

Some people see letters and numbers as different colours, a syndrome known as grapherne colour synesthesia. Everybody's colours are different, although apparently red is the most common hue for A. Characters are tinted or shaded with colour rather than appearing to be completely green, orange or whatever. Synesthetes can find that this syndrome helps with spelling. For example, they might know a word starts with a K because it is green.

# CHUNKING INFORMATION

We have seen that grouping information often comes naturally and can help us remember more in a meaningful and useful way. The technique of breaking down the information you want to learn into smaller sections is called chunking. We've already seen an example of it in the ABC song described on page 92 in which the alphabet is recited in small groups of letters.

## Word chunking

When you are planning what you will do during the day, it makes sense to group your appointments and 'must do's' into parts or times of the day. For example, in the first half of the morning you might have a meeting and want to get started on a high-priority task. Between then and lunch you may need to prepare some figures for a meeting that afternoon. Your lunchtime priority may be to catch up with a friend. In the afternoon you have another meeting and need to check progress of a colleague's project, before setting up a presentation for the next morning. In the early evening you want to buy some flowers and the vegetables to go with your evening meal, and so on. You have 'chunked' what you need to do into time bands.

Putting words into categories, whether it be nouns, verbs and adjectives, different types of shopping, or whatever, is a 'word chunking' technique. It becomes easy with practice because words have meanings, which makes it straightforward to identify categories for them. Numbers are different, because they don't have meaning for us – but chunking is a very valuable tool for helping us remember strings of numbers.

## Number chunking

What's the pin number for your bank card? What about your credit card?

Can you remember long sequences such as your bank account number? How about your best friend's phone number?

Recalling numbers is part of everyday life. OK, you can store numbers on a phone, but in an emergency where you had to borrow someone else's mobile to call a friend, would you know what to dial?

We can store on average seven items in our short-term memory (plus or minus two). So in theory we can remember a seven-digit number such as 6087149 (try it!). However we can make this task much easier by grouping the numbers into chunks: 608-71-49. Now there are only three, larger, numbers to remember.

> ### Start chunking
>
> Learn the ten-digit number **2546916287** as **254-691-6287**. Study the three large numbers, saying them to yourself (it often helps to give the numbers a rhythm). Test yourself on your recall.

Some numbers arrive almost 'ready chunked'. For example, landline telephone numbers are often presented in two chunks: the STD code followed by the remaining six digits, which can be chunked into two groups of three.

> ### STD + 2 x 3
>
> Try learning the landline telephone numbers of two friends by chunking them into the STD code and two sets of three.

Similarly, mobile telephone numbers always start with 07 (so you will always start with the 'chunk' 07) followed by nine digits which can be chunked into three groups of three. Chunking in same-size groups allows you to use rhythm in remembering them.

## Mobile memory

Try learning two mobile telephone numbers in this way. You will probably get into the habit of remembering phone numbers using the same chunking pattern. It can be disconcerting when someone tells you a number using a different pattern, because you have to convert this to your preferred system.

## Pairs to patterns

Get someone to read you a telephone number that you are unfamiliar with in pairs of numbers which you must then convert to your preferred pattern and learn.

Other multi-digit numbers we come across in our everyday lives are credit and debit cards that are 16 digits long, usually presented on the card as four groups of four.

> ### Chunks of memory
>
> If you don't know it already, learn your main debit or credit card number, deciding whether to use the four-digit chunks or to break it down another way.

## Studying using mnemonics and chunking

These two memory aids can significantly boost your study skills, from learning to revising and performance in tests. As with any data to be memorised, it is very important that the initial input is accurate: don't learn things wrong! Error-laden information can become enmeshed in your memory as part of a web of associations that can be very difficult to unravel.

When you are trying to learn and remember how something works, take a little time to generate your own example, linking it to an activity you enjoy. For example, an engineer who needs to memorise a certain ratio may be a keen cook. They could picture how to use the ratio to change quantities of food in a recipe so that it feeds a different number of people. Such comparisons not only build understanding of what you are learning, but help to create further associations, making the information more memorable. These links do not have to be realistic, either, indeed it might help to make them ridiculous and fun. For example, if the engineer was also using other images to help with their studies (say, colours and animals), they could imagine using the ratio to provide blue party food for a group of penguins (see advice on creating images on page 114).

## Time chunking

You need to be in the right frame of mind to study effectively. This means getting rid of distractions (turn off the phone and don't work next to the fridge – every time your attention is diverted it takes up to 15 minutes to regain your full concentration). Sleep well and eat foods that release their energy-giving sugars slowly – porridge for breakfast is a good example. This should help to even out the inevitable peaks and troughs in attention span.

Plan how you will study: hours staring at a book are not as productive as intense shifts of 20 to 50 minutes followed by reviewing what you have learned and, importantly, relaxation. You should also plan further reviews of your learning after a day, a week and a month. You will want to review as briefly as possible and that is where mnemonics come in: condensing information to a small number of triggers which lead to fuller content.

## Two study methods

Two widely-used study methods have their own mnemonics: **SQ3R** and **PQRST**.

**SQ3R** stands for **Survey**, **Question**, **Read**, **Recite** and **Review**. The five-step method can be used for any kind of learning. It was developed for rapid training of military staff during World War II and has since been adopted by many educational institutions. In summary, it works like this:

- ✔ **Survey:** get a general overview of the topic, perhaps by studying the contents page of a book (or a chapter overview) on the topic or course material learning objectives. This should take no more than five minutes.
- ✔ **Question:** Repeat, but think of questions that you hope will be answered by the material – you can do this by turning the existing headings into questions. This builds interest and is the foundation for later associations.
- ✔ **Read:** Do this without taking notes.
- ✔ **Recite:** Answer the questions you set. You don't have to say them out loud (unless it helps – some people whisper it). Make notes if you wish.
- ✔ **Review:** Go over what you have learned, checking what you have taken in against your questions and identifying any weak areas.

A similar study method is called **PQRST**. This description includes some of the variations in the words that form its mnemonic.

- ✔ **Preview:** Scan the material to identify key elements.
- ✔ **Question:** What do you already know? What do you want to know?
- ✔ **Read:** At your own pace.
- ✔ **Study/State/Summary:** Make notes of the key information.
- ✔ **Test/Turn back:** Would you be ready for a test on it? If in doubt, Turn back and go over what you are weak on.

### Study letter

Learn the meaning behind the acronym **SQ3R** or **PQRST**. Explain it to a friend or colleague.

Use this process for each topic you are studying, so that your learning is broken up into bite-size chunks.

## Making notes

Making notes is a skill in itself. Notes should be as brief and meaningful as possible – maybe just a few words on the content of each page. If these can be chunked into categories and summarised into mnemonics or acronyms, so much the better. Start by using a highlighter pen to pick out the key facts and phrases. Condense these into a short set of notes – it doesn't have to make sense to anybody else, so you can happily ignore the rules of grammar and spelling. When you review these you may be able to summarise them into phrases or simple mnemonics.

## Why cramming doesn't work

The time-honoured practice of students spending a whole night trying to cram in the contents of a course after partying away the term might earn a few marks but it is unlikely to store information in your long-term memory. In addition, the stress induced by having to cram, and the sleep lost through doing it and worrying about it, are likely to reduce what you can remember. If you are lucky, you might succeed in boosting your score in an exam, but you haven't added to your store of skills or knowledge, and if your course is vocational, you haven't gained real knowledge.

# SHORT CUTS

This section shows you how to create links and associations as you study information so that you can remember more. We make associations all the time: for example, the sayings 'bread and butter' or 'raining cats and dogs' link pairs of words. You can train yourself to produce associations at great speed – the key is practice. This section outlines a number of systems you could use. It makes sense to consider how well they may work for you, and try out a few of the exercises before committing yourself to just one. The one that is likely to work best for you is the one that reflects your auditory/visual/sensory preferences (see pages 42–74).

When you become accomplished at creating images in many different ways, from actions to wordplay, you need to ensure that these valuable associations are stored in your long-term memory. One technique for achieving this is called expanded rehearsal and is explained in this section. The next stage is to link these images together so that you can learn information in sequences – this is the basis of the methods used by memory champions to recall massive amounts of data in order.

## CREATING IMAGES

Creating images to go with words or ideas is a skill. It will come more naturally to some people than others, but everyone can improve at it. Close your eyes when creating and (especially) retrieving images as this gives your imagination free rein: brains are good at conjuring up images out of nothing, which explains the popularity of stories and plays on the radio. This series of exercises is designed to show how some things are easier to visualise than others.

> ### Don't touch
>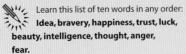
> Learn this list of ten words in any order:
> **Idea, bravery, happiness, trust, luck, beauty, intelligence, thought, anger, fear.**

Most people would find this task very difficult because all these words are abstract nouns: feelings or qualities or things we understand but can't touch or picture readily.

### Handle with care

Learn this list of ten words in any order:
**Cloud, cat, dog, volcano, chair, drum, spade, kite, ball, plane.**

That should have been much easier, because these are all concrete nouns: things we can see and touch and which we can readily form into a mental picture.

The challenge for the memory improver is to be able to conjure up memorable images quickly, even of things that aren't 'visual'.

### Reach out

Try the first exercise again, concentrating on making up a simple but clear image. For example, 'idea' can be represented by a bulb over a head, 'bravery' by a hero such as Batman or Superman, 'beauty' by a pretty face, and so on.

Bearing in mind that memory is multi-sensory, you may find it helpful to make your image an action. Language teachers do this a lot because acting out, say, 'swim' while learning the word in a new language creates an action trigger. The next activity uses ten

Polish words. If you speak it already, find ten words from a language you don't know.

> ## Physical Polish
>
>  Learn these ten words by performing a physical action – the simplest gesture that occurs to you - as you repeat the word. The last two will probably present more of a challenge than the others. Get someone to test you.
>
> | | |
> |---|---|
> | **One:** jeden | **Eat:** jesc |
> | **Two:** dwa | **Ball:** pocisk |
> | **Swim:** plywac | **Cat:** kot |
> | **Jump:** skok | **Cinema:** kino |
> | **Smile:** usmiech | **Flower:** kwiatek |

If you found the gestures useful, build them into other memory images you create – you may be able to refine your technique so that just the thought of the movement acts as a trigger, or you may end up waving your arms around a lot as you recall things!

## Using sounds and puns

The sound of a word or a pun related to it can be a terrific memory trigger. For example, Canada starts with can, so you could picture a soft drink can. Porcelain figures can be made from (and in) China.

## Sound triggers

Learn this list of 12 countries in any order, creating sound triggers. If it helps you may like to add your own sound images to these. There are ideas for these after each word, but you might find a better one that works for you.

**Canada** (a drinks can, or can aider: the clang as someone puts a can in the recycling bank).
**China** (a figurine made of china. Imagine the sound it would make if you were to tap it).
**Fiji** (fidget, someone jumping about).
**Germany** (reverse the words 'germ' and 'many' to make 'many germs').
**Greece** (grease. You could think of an image of grease being used on a machine or greasy food).
**Hungary** (hungry. Someone rubbing their stomach).
**Jamaica** (jam-maker. You could imagine a Jamaican person standing over a bubbling pot of jam).
**Luxembourg** (luxury gives the first sound).
**Panama** (the famous panama hat).
**Russia** ('Rush ya', from the old joke: 'I don't want to rush ya, but if you must go [Moscow] you must go').
**Syria** (syrup. Imagine the glugging sound made when pouring the thick liquid).
**Turkey** (bird that goes gobble-gobble).

## Strange route

Practise creating truly memorable images. For many people, these are likely to combine two unlikely elements or actions. For example, to link Turkey and Syria, take your turkey and syrup pictures and sounds and imagine the poor old bird going 'gobble-gobble', sitting in a hot oven while it contentedly bastes itself with syrup. Hear the sound of the thick liquid sizzling as it touches the skin. Exploit the multi-sensory nature of memory by adding some cooking smells or taste the turkey meat and the sweet syrup. It may also help to have some movement in your image, making it more active. Maybe the turkey could be looking round for more syrup, or flapping its wings.

Of course we have strayed a long way into unreality here, but the point is to make your images vivid. Don't be afraid to let your imagination run riot. If the image is funny, so much the better. However, some people prefer logical, more rational combinations. They might prefer to see a frozen turkey next to a can of syrup on a supermarket shelf or in the kitchen. The turkey might just have a syrup container tucked under its wing in the farmyard. If this is the case with you, you may enjoy the story method (see pages 177–8), with its reliance on some kind of logical narrative.

# LINK IT UP

On page 120 is a list of countries, which is the basis for the next stage in developing a better memory: linking two images together. The suggested word-play reminder for Russia was an old joke: 'I don't want to rush ya, but if you must go you must go' in which 'must go' is pronounced as in Russia's capital city, Moscow. The phrase is memorable because it is funny and it provides a link between the country and its capital.

If you found yourself remembering the phrases rather than creating a picture, that is fine if it works for you. However, the blending of ideas to create one image with two bits of related information is a vital memory-building technique because it doubles what you can learn. Just as supermarkets have bogofs (buy one, get one free) so you have a rotgot: remember one thing, get one free.

The saying 'a picture is worth a thousand words' is relevant here: we can continue adding new things to a picture in a way that would turn what begins as a snappy phrase into a long-winded paragraph. Our brains have the capacity to store about 10,000 images – but they would find it hard to cope with 10,000 little sayings, however concise they are. So do keep trying to make images rather than sticking with words.

## Sound links

Use the information in the table to make links between the 12 countries and their capitals. There are suggestions for ways to create a combined image: feel free to make up your own. Get someone to test you on your recall.

| Country | Capital | Image |
| --- | --- | --- |
| Canada | Ottawa | Otter swimming away ('otter-away') with a can. |
| China | Beijing | Figurine in a beige colour. |
| Fiji | Suva | A fidgeting saver jingling money in their pockets. |
| Germany | Berlin | Burly germs – really big, heavy bacteria. |
| Greece | Athens | Throwing grease at hens. |
| Hungary | Budapest | An annoying Buddha (Buda-pest) pointing to his rumbling stomach. |
| Jamaica | Kingston | The royal (king) jam-maker: a cook wearing a crown. |
| Luxembourg | Luxembourg | An easy one! |
| Panama | Panama City | A city in a hat. |
| Russia | Moscow | 'I don't want to rush ya, but...' |
| Syria | Damascus | A river of syrup blocked by a dam. |
| Turkey | Ankara | A turkey in a tank (T Ankara). |

## Pairing up

 Find other information pairs – you could continue with countries and capitals, or switch to number-one hit singles and their performers, artists and pictures, or any facts that you are keen to learn, perhaps relating to a hobby or your studies. When you have ten pairs, create images where the two facts are combined and learn them. Test yourself on how successfully you can recall all of them.

## Using images to put names to faces

Creating mental images can be a powerful way of putting names to faces. For example if you meet a Theresa Key, imagine a key in the middle of her face. Chris Street could have a roadway down his nose, while you could picture Sarah Wise as a wise owl with big round glasses. George Bush could have a hedge growing out of his mouth and Madonna could be holding a baby. Incidentally if you like the idea of using famous names, check out the Dominic method on pages 174–6. This works for first names, but of course there is the problem that you will meet many Marks (with those distinctive blotches you've pictured on their faces) and Petes (with their peaty-brown hair).

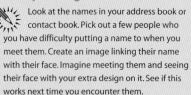

### Face painting

Look at the names in your address book or contact book. Pick out a few people who you have difficulty putting a name to when you meet them. Create an image linking their name with their face. Imagine meeting them and seeing their face with your extra design on it. See if this works next time you encounter them.

Another way of creating an image is the self-reference technique. Imagine you meet someone with the same name as your partner. Immediately picture them shaking hands or squashed up together. This is a pictorial way of 'filing' people with the same name together. The link does not have to be via the name: if you have a file for 'financial people' and you meet someone who works for the bank, you can picture them all in the bank welcoming the new member of the gang.

This method can work for appointments too. Imagine you have an important meeting scheduled for the same day as your father's birthday. Picture him blowing out the candle on his birthday cake as you head for the door to travel to the appointment.

# Imaging and studying

Just as it is important to learn 'error-free' so that you don't have to go back and unlearn things, so it can be vital to study actively, rather than just read and condense. This means getting involved in the subject and thinking and talking about it. Having a study partner can be useful: discussing what you have learned, making links with what you already know will build good understanding and secure memory. One way of checking your recall and comprehension is to 'teach' a topic – either for real to a colleague, or as a role play.

## Two-minute talk

Learn about something new that you are interested in or that relates to your job or hobby. This can be anything from how clouds form, what the colour wheel is, the rules of netball, a foreign language, etc. Make a two-minute presentation to someone else about the topic, summarising it into no more than five points. Illustrate these with vivid images that bring the subject to life – perhaps just by explaining how you have remembered them. If at all possible, make your presentation from memory with only the briefest of notes. We always talk more fluently when working from bullet points rather than reading out a prepared text – and it is much easier to listen to someone speaking 'off the cuff'.

By planning what you would say and how you would say it, you are reinforcing your learning, deciding how it is relevant to you (self-referencing) and working out how to make it interesting and accessible to others – which calls for images and associations that in turn will aid your own recall in the future.

## Expanded rehearsal

This is a technique for ensuring information gets into your long-term memory. It works for any memory but it is particularly worth pursuing with something you really want to learn and will use in the future. Using it with images instead of individual facts means that you can remember more from the same amount of effort. The process is illustrated in the flow chart opposite.

By this stage that set of information should be locked in your brain. This may seem fussy and prescriptive, but the tricky part is remembering to review: the actual review occupies a matter of minutes. The point is that your brain is practising retrieving and interpreting images and, like anything else, the more you practise the easier it gets.

This regular revisiting also ensures that the memories do not get filtered out from a short-term depository. Get into the habit of reviewing on train journeys, at the bus stop, in the supermarket checkout queue and at all those other times when you have a small amount of spare time. The combination of information-packed images and expanded rehearsal is a powerful memory tool.

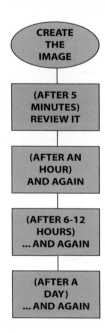

# REMEMBERING THINGS IN ORDER

The double image technique also takes us a step nearer the methods that allow remembering of vast blocks of information in section four. However, it is also very handy for recalling information in sequence.

## Peg system

The number rhyme, or peg, system is an extremely effective system for remembering (initially) short lists of information in order by combining two images. In this, each number up to ten is given a rhyming word partner which is very easy to visualise so that you use pictures in your mind to link the number with the data to that you can remember it in order.

It is up to you to choose rhyming items for each number, but here is a list of commonly used options. Notice that the rhyming words are all nouns, which are easier to visualise than more abstract words such as verbs or adjectives:

✔ **One** = bun, nun, sun
✔ **Two** = shoe, glue, loo
✔ **Three** = tree, bee, knee
✔ **Four** = door, saw, paw
✔ **Five** = hive, dive

- ✔ **Six** = bricks, sticks
- ✔ **Seven** = heaven, Devon
- ✔ **Eight** = date, skate, gate
- ✔ **Nine** = line, vine, sign
- ✔ **Ten** = hen, men, pen

You may find it useful to add **Zero** = hero

**Fee: bun guinea**
The number rhyme system (also known as a peg system) was invented around 1880 by John Sambrook. In 1885, he was offering to teach it privately for a fee of one guinea, or by correspondence for 10s. 6d. – an example of the enormous interest in memory systems in the late Victorian era.

> **Bun shoe tree**
>
> Create and learn number rhymes for the numbers up to ten. Practise keeping the images consistent (draw them if it helps). Practise picturing them in order, and out of sequence. If some associations are weak, change them to something with a stronger, more memorable picture.

### Quick thinking

Practise using your number rhymes to make rapid associations with door numbers, pin numbers and other short numbers. Move on to memorising a telephone number using this method.

### Try it out

Practise using the number rhyme system making instant associations between the number rhyme and its object to memorise this list, in order. Get someone to test you on finding the seventh, third, tenth item.

**List: 1. computer. 2. lamp. 3. book. 4. scissors. 5. light bulb. 6. paperclip. 7. birthday card. 8. mouse mat. 9. radio. 10. photograph.**

The number rhyme system can be used to remember number facts by combining two digits into one image (just make sure you get them the right way round).

**One is a chew**

If you use your auditory memory a lot, you may find it useful to attach sounds to your number rhymes. One could be the sound of the bun being chewed. Two could be a shoe squelching through glue. Three would be the buzz of a bee and four a slamming door. If you find your other senses are good memory triggers, use those instead – smell might be worth a try.

## Zero is a ball

An alternative (or an extension, if combined) to the number rhyme technique is the number shape system, in which you give numbers a pictorial mnemonic that shares its basic shape. This method may suit a strongly visual learner who readily imagines shapes. You should make your own shapes that are meaningful for you so that they are vivid – by all means include other senses if it helps – but do make sure that none can be confused with other images. Here are some ideas:

**0** is anything round, such as a ball, or a tyre.
**1** is a line or anything straight, such as a candle or a pencil.
**2** is a curve like a swan's neck or a poised snake.
**3** is a bird or a butterfly.

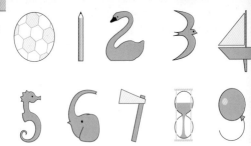

- **4** is a sail on a boat, or a flag.
- **5** is a hook or a seahorse.
- **6** is a lasso or an elephant's trunk.
- **7** is a boomerang or an axe.
- **8** is an hourglass or a snowman.
- **9** is a balloon with trailing string or a flag (if you didn't have a flag for 4).

You may find it helpful to consider images that are related by concept rather than shape: for example, 4 could be a car (four wheels) and 5 a glove (five fingers and letters), and perhaps 9 can be a cat (nine lives). As always with these techniques, your own preferences are more important than the approaches other people use, because they will help you make things memorable for you and how your brain operates. What does your pin number look like?

### Draw it out

Create your own shape images for the numbers zero to ten, drawing them if it helps.

### Pictures of ten

When these images are secure, practise using them to memorise this list of ten objects in sequence.

**1** hammer **2** clipboard **3** hat **4** window
**5** bird **6** bowl **7** blossom **8** cliff
**9** envelope **10** lamppost

## Remembering 20 items in sequence

You can combine the number rhyme and the number shape systems into a strategy for remembering 20 items in order: just decide which of them will represent the numbers 11 to 20. For example, one can be a bun but if you are using rhymes to represent numbers above ten, it can just as easily stand for eleven (see sequencing exercise on page 132).

## Remembering numbers up to 100

The number rhyme system can be expanded to remember longer lists of up to 100 numbers. As we have seen, you can remember up to ten items by

> ### Sequencing
>
>  Study this list of 20 items, attaching a number rhyme or number shape to each one.
>
> | | | | |
> |---|---|---|---|
> | **1** spade | **2** gate | **3** glove | **4** oven |
> | **5** battery | **6** magnet | **7** dice | **8** plane |
> | **9** diamond | **10** car | **11** string | **12** soap |
> | **13** wardrobe | **14** stone | **15** bell | **16** belt |
> | **17** apple | **18** pill | **19** coin | **20** ladder |
>
> When your images are in place, list the items in order, then get someone to ask for them by position in a random order.

creating an image linking number and object. Since the digits zero to nine must be repeated in all numbers over ten, and appear in the same order when we count up in sequence, they can be re-used to complete larger numbers if there is a clear system for describing which set of tens the number is in. So you need nine 'grouping' images. You could use colours, feelings or any other association that is vivid and memorable, as suggested here for the first five groups:

✔ **Numbers 10 to 19:** red, on fire.
✔ **Numbers 20 to 29:** blue, covered in water.
✔ **Numbers 30 to 39:** white, frozen or on a cloud.
✔ **Numbers 40 to 49:** green, in a grassy field.
✔ **Numbers 50 to 59:** yellow, covered in egg yolk.

And so on to 99. Under this system, 11 would be a bright red or blazing bun, while 21 would be represented by a bun floating in a pool, and 31 could see the bun being put in the freezer. The bun would be on a cow's horns in a green field for 41, and served with an egg for 51. This is called the expanded number rhyme system.

### Through the decades

Choose memorable general settings for numbers in the tens, twenties, thirties, forties and fifties, either using the suggestions above or, even better, making up your own. Think of images for numbers up to 59. Get someone to write and say a random sequence of ten numbers under 60 and learn it using this system. When you've done this a few times, move on to the next exercise.

### Page turner

Study the first 20 pages of a magazine on which each page has at least one vivid image (comic books are ideal). Learn to associate the number image with the contents of the first 20 pages so that you can describe one feature of each page, identifying it by number. Get someone to test you by calling out page numbers at random.

## More pages

Repeat this activity with 30 pages.
See if you can build up to being able to
describe the first 50 pages.

The following exercises use a deck of cards to
provide a list of 52 items (cards) to be remembered
in sequence. This is harder than the previous
exercises as you will need to conjure up an image or
association for each card. For advice on card images,
see pages 170–73.

## Card sequence

Shuffle a deck of cards. Turn the first 20 over
one at a time, creating an image linking the
card with its place in the sequence. When you can do
this proficiently, move to the next exercise.

## Deck sequence

Add ten cards at a time until you can memorise
an entire deck in sequence using the expanded
number rhyme system.

# AS SIMPLE AS ABC

An alternative to using numbers for remembering sequences is to go back to the good old alphabet that most of us can recall rapidly in order. You can allocate each letter a rhyming word that also carries a powerful image. This is a phonic/image system so it is vital that you choose words that rhyme with the letter – they don't even have to start with it. For example, 'apple' or 'axe' wouldn't work as well for 'A' as 'alien' or 'ace' because you need the long, not the short, 'a' sound, while 'entrance' works for 'n'. Here are some suggestions for alphabet system images.

- **A** ace or ape
- **B** bee or bean
- **C** sea or seed
- **D** diesel or dean
- **E** east or eagle
- **F** effigy or effort
- **G** jeans or gee-gee
- **H** H-Bomb or hate
- **I** eye or ice
- **J** jade or jay (bird)
- **K** cake or cane
- **L** elastic or elbow
- **M** empty or embers
- **N** entrance or envelope
- **O** oak or oboe
- **P** pea or peep
- **Q** queue or cute
- **R** arc or arch
- **S** escalator or essay
- **T** tea or team
- **U** ewe or unicycle
- **V** vehicle or veal
- **W** wall or WC
- **X** X-ray or ex
- **Y** y-fronts or wire
- **Z** zebra or zoo

As with the other methods, you create a mental image linking the word-letter with the thing you want to remember. This will probably require more practice than the number shape system, but some people (especially those who think visually and not in words) may find it works better for them.

## First half

Learn an image for each letter from the first half of the alphabet: the letters A–M. Use it to remember these sports items in sequence:

| | | | |
|---|---|---|---|
| **1** ball | **2** stick | **3** boot | **4** net |
| **5** racket | **6** umpire | **7** goal | **8** referee |
| **9** pitch | **10** whistle | **11** court | **12** kit |
| **13** bat | | | |

When you are confident you can list these items in and out of sequence, move on to the next activity.

## Second half

Learn an image for each letter from the second half of the alphabet: the letters N–Z. Use it to remember these sports items in sequence:

| | | | |
|---|---|---|---|
| **14** stumps | **15** hat trick | **16** catch | **17** crowd |
| **18** player | **19** cap | **20** spectator | **21** studs |
| **22** manager | **23** pass | **24** track | **25** baton |
| **26** team | | | |

## The whole game

When you are confident you can list these items in and out of sequence, you can combine both lists. Practice recalling the two lists put together, numbered 1–26, in and out of order. Get someone to test you on this.

## Extra time

Add the following 26 leisure terms to the list you have already learned, linking each to a letter of your 'second alphabet' in some way:

| | | | |
|---|---|---|---|
| **1** ipod | **2** cinema | **3** ticket | **4** game |
| **5** restaurant | **6** golf | **7** concert | **8** painting |
| **9** gym | **10** DVD | **11** backpack | **12** music |
| **13** museum | **14** rugby | **15** dancing | **16** swimming |
| **17** aerobics | **18** chess | **19** tennis | **20** karate |
| **21** walk | **22** cards | **23** quiz | **24** party |
| **25** book | **26** tent | | |

You can expand this method by creating a 'second alphabet'. This repeats the alphabet but all the images second time around must be instantly identifiable as being in the second series. You could do this by putting a number two somewhere in the image, or by using a specific colour, so that, say, all 'second alphabet' images are predominantly blue.

# CHAINING MEMORIES TOGETHER

When we recite the alphabet or count in ones to 100, we don't hesitate before naming the next 'fact' in the list: we are so familiar with it that each item triggers the next. This is chaining or linking: a series of memories each triggering the next in the sequence (A triggers B, which triggers C, and so on). It is a fine way to remember long sets of items in order. The trick is to turn the data into paired facts: for example AB, then BC, then CD. This pairing is achieved by creating visual associations.

Here is how the system could work for memorising six random words: kerb, rug, fossil, island, nut, and tent. The first two words can be paired by imagining a brightly patterned, tasselled rug curved over the side of the road (or maybe just the kerb woven into a rug). The next pair could be a rug wrapped into a roll with a dinosaur bone (the fossil) sticking out of the end. Then you could picture a fossilised island: palm trees and sand all turned to solid rock, or maybe just a fossil-like rock floating in the sea (forming an island). Island and nut is an easy pairing of a cartoon island with a nut hanging from its single palm tree. For nut and tent, imagine a metal nut on top of a tent pole.

Of course it doesn't matter what the images chosen are, so long as you remember them. If you enjoy puns and wordplay, your images are likely to feature this; if you are a very literal person, you might just picture the two items squashed up against each other.

> ### Paired images
>
> Learn these randomly selected items in order by putting them into paired images. Get someone to test you on them, in sequence, and by asking for the fourth or sixth item.
>
> **1** violin  **2** magnet  **3** rifle  **4** fly
> **5** cone  **6** helicopter  **7** potato  **8** bicycle

If any of the pairs are not easy to remember, close your eyes and try concentrating on the image you made for two to three seconds, and see if that improves your recall. If it does, you need to practise totally immersing yourself in making your visual links, possibly with your eyes shut. If it doesn't, maybe the visual link wasn't right for you. Was it too bizarre or, conversely, not strange (and therefore memorable) enough?

> **More pairs**
>
> Get someone else to create a list of eight random objects (not things in the room, and not too abstract). Ask them to say each item to you at five second intervals. You are not allowed to ask for any to be repeated. As they do this, concentrate on creating your paired image. Try not to worry about whether you'll remember the first or the middle ones as this will simply distract you. See how well you can recall the list in order.

A disadvantage of the link system is that forgetting one link probably limits your capacity to reach the rest of the items in the chain. If this occurs, keep practising, but consider trying one of the other methods for sequencing memories instead.

### Concept maps

The concept map is another way of linking images and information. This is a form of note-taking that exploits our liking for visual information. Write the main subject in the centre of a piece of paper and draw lines away from it to create 'branches' for themed information. This is a way of organising notes by category, forcing you to use the minimum of words or pictures. Students who use this system can often recall the overall shape of the 'map' they created and can travel along its branches to find their triggers.

# SPRINTING

## Introduction

This section deals with advanced memory techniques that world memory champions use. Some of these techniques date back thousands of years and have proven value. It takes a lot of work to become proficient in any of these methods, but they are effective. Read through the suggested techniques and try the exercises to get an idea of which methods might suit you. The competition website is **www.worldmemorychampionship.com**

### Memory champions

The annual world memory championships celebrate the pinnacle of achievements in memorising facts and figures. Some events (and the current records) are:

| Event | Record |
|---|---|
| ✔ Memorise a single pack of playing cards as fast as possible | 31.16 seconds |
| ✔ Memorise as many packs of playing cards as possible in one hour | 27 packs |
| ✔ Memorise a list of random words in 15 minutes | 214 |

## Using routes

Imagine yourself holding an onion in your left hand, with a carrot hanging from your elbow and an apple resting on your shoulder. In the same locations on the other arm are a loaf of bread, a jar of coffee and a bar of chocolate. This is the body system: the imaginary placing of items on parts of your body. It is a good way to start to learn how to use routes to memorise things.

### Body art

**List:**

| | |
|---|---|
| onions | carrots |
| bananas | apples |
| strawberries | bread |
| coffee | chocolate |
| washing powder | batteries |

Adapt the shopping list to things that you actually buy, then imagine placing each item on part of your body. Make it funny by hanging bananas from your ears and wearing the box of washing powder like a shoe. If it helps, put similar items near each other – hang all the fruit on your left arm, for example. Then try it out when you go shopping.

This method is kinaesthetic at least to the extent that you are picturing the items on your body and can imagine how they would feel and look, so it may suit practical people who learn by doing. It is limited, of course, to how many items you can 'place' around your body. This figure is likely to be well under 20. If you are good at visualising where things are, and perhaps have a strong sense of direction, the following method may suit you better.

Write a shopping list in the order in which you will find each item, by taking your usual route round the supermarket (we are encouraged to travel from the left to the right of the store), or by following the layout of your wine shop, or the order of shops in the high street.

### Route march

Add to the shopping list above so that you have about 15 items. Write them down in the order you will encounter them in the store. Picture yourself moving between each one (making links as on page 142) and make the images as surreal as you wish. Leave the list at home and see if you manage to buy every item using your sequenced images.

## Location, location, location

We find it pretty easy to visualise walking round a shop and being aware of what will be stored where (this is why some supermarkets re-jig their displays, so that you will chance across items that you normally miss). Check how well you can visualise your own home, because this will be the basis of a key memorising technique: the Loci or Roman room technique. Strictly speaking, this involves creating one, large, imaginary room in which to store images. Roman writer, Cicero explained the requirements: 'One must employ a large number of places which must be well-lighted, clearly set out in order, at moderate intervals apart, and images which are active, which are sharply defined, unusual, and which have the power of speedily encountering and penetrating the mind.' By all means experiment with this, but it is easiest to start by using the place with which you are most familiar: the home where you live.

### Home visit

Take a mental walk around your home. In 'real time', imagine walking through the door to each room, studying what is there, and leaving to enter the next room. Think of two things in each room that usually catch your eye and note them down. Now go and check that they are where you thought they were.

Walking around the home is a familiar journey that we can mentally reproduce without much effort. The Loci or Roman room method (also known as the memory palace) exploits this familiarity to aid the memory: you are going to imagine things from a given list in each room, so that by touring your house you can remember each item. If you prefer, you can create an entirely imaginary house with memorable features such as pillars, murals, art masterpieces and sculptures. Naturally, the key thing is to remember each of these elements, as they are the 'pegs' on which you'll hang images.

## Rhino on the sofa

Repeat your mental tour and identify ten specific places in your house (remember, it can be a complete fiction if you prefer), each in a different room or at least not near each other (you can make use of stairs, corridors and hallways if you wish). Learn this list of exotic animals in sequence by imagining yourself placing each item in the next location in the tour:

**1** crocodile    **2** giraffe    **3** ant eater
**4** meerkat    **5** elephant   **6** rhinoceros
**7** monkey    **8** snake      **9** flamingo
**10** lion

The choice of zoo animals rather than household items should make their presence in your 'house' more memorable. Try to picture the animals as realistically as possible, not as if they are stuffed toys. So have the monkey climbing the curtains and the giraffe's head poking through a window or skylight. As has been noted, vivid images are more memorable than everyday ones. This leads us to…

### Think big

You are going to use the same technique to memorise a list of more common items. To make them stand out from the room in which they must be placed to keep the sequence correct, make the image bizarre: a giant paperclip, a coat floating through the air, a ball-shaped television (for remembering ball, not the TV!) or make other objects change colour, burn or move.

**1** paperclip **2** coat **3** ball **4** sausage **5** orange
**6** DVD **7** hat **8** radio **9** photograph **10** box

You should be able to achieve this if your images are sufficiently memorable. If they weren't, you need to concentrate on creating outlandish pictures that will imprint themselves on your memory.

# The origins of the room system

The room system was apparently first suggested by a Greek poet called Simonides in about 500 BC. He was asked to give a recitation in honour of a wrestler hosting a party to celebrate success in the Olympic Games. Having delivered his eulogy, Simonides was called outside. Just then the building collapsed, killing the remaining guests and crushing their bodies beyond recognition. The poet was able to help the victims' relatives to identify the corpses through remembering where he saw people sitting in the room just before he left.

## 'In the first place…'

The room system became popular with classical orators such as Cicero, who prided themselves on speaking without notes. This is the origin of the term 'In the first place… In the second place…' and so on, as the speaker moved mentally around the room picking up on the following point at the next place.

The Loci system has an advantage over the paired association chain system described on pages 150–51. If you forget one image, you don't lose the rest of the chain, because you can still 'move on' to the next room. Of course, the limitation in the system as described so far is that you only put one item in each room. However, once you are using the room method confidently, you can start to remember longer lists by placing more objects in each room.

## Double rooms

Repeat your house tour. In each room, identify where you could place two objects, preferably far apart. Try this mentally, then walk it for real to check. Try starting your mental tour at different points, recalling the pair of locations in each room. It may help to place them at different levels (such as hanging from the ceiling or under a chair). Write, sketch or recite your two locations for each room.

## Melted biscuit

Memorise this list of 20 items in order using this method:

| | | |
|---|---|---|
| **1** biscuit | **2** submarine | **3** drum |
| **4** parrot | **5** ladder | **6** flag |
| **7** pen | **8** bullet | **9** cigarette |
| **10** ship | **11** teddy | **12** bell |
| **13** pyramid | **14** tree | **15** throne |
| **16** helmet | **17** toothbrush | **18** compass |
| **19** cake | **20** sword | |

The list above includes everyday household items as well as things you would not expect to find in the home. When you have completed the exercise, analyse which ones were easier to place and recall. The trick with the mundane objects is to give them strange characteristics such as being massive, misshapen or melted – something to make you register them. Try different starting points in your house journey and identify any weak points where you get stuck. Some people use this technique to remember people they have just met.

> ### Common people
> Write a list of 20 people who are acquaintances rather than good friends – a set of work colleagues or a class of fellow students, for example. Memorise the list using the room system. Again, analyse the results to identify any weak points in your house route, and whether there is a pattern to any mistakes (gender? characteristics? how you feel about them?).

# Loci through history

The Loci method survived from the ancient world and was used in medieval and Renaissance periods as a way of reading, remembering and meditating about the Bible – and was also used to aid composition. It was apparently widely taught to children until 1584, when it came under attack from the Puritans who objected to its use of outlandish thoughts (because as we know these are far more memorable than banal everyday things).

You can continue to add 'trigger locations' to your imaginary tour, placing items on or near chairs, tables and in nooks and crannies. After all, every room has four walls, four corners, a ceiling and various other possible locations. There is a danger that the route will become overloaded and that you will start to confuse items – to which the solution is the 'journey route' described later on (see pages 154–9). However, it is worth experimenting to see if you can build up to 40 and then 100 locations using your home.

## Four sides

Go on your house tour, identifying where you could place four objects in each room or area, aiming for 40–50 locations. Practise recalling each location, in sequence, as in the exercise on page 148.

Because you are going on a familiar route, and provided you always look around the room in the same order (perhaps scanning the ceiling, then the walls from left to right, then the floor area), you can remember sequences.

### Using Loci

Create your own list of 40 items to be memorised in order and learn this sequence using the Loci method.

If you coped well with that, move on. If not, consider whether it is worth going over the method a few times, or consider the other techniques in this section. The next activity is likely to stretch you, as the information is more abstract.

### Too many lists?

One natural concern when using the room method repeatedly is the worry that separate sets of information will get mixed up. This rarely happens provided you leave at least a couple of days before using the system for new data. However, another option is to deliberately combine lists as you make them, so that, for example, each location stores three items. Provided you remember them in order (list one, list two, etc. – colours may help) you can select from the right set of information at each location.

## Shakespeare's order

Study the list of the plays of William Shakespeare in the approximate order in which they were written. You may not be surprised to learn that the exact order is hotly debated by academics. There are 38 plays on the list. Learn them in order, using the Loci method based on your home. Clearly numbers will help when a series of works was written. As always, make your images as vivid as possible.

**Henry VI, Part I**
**Henry VI, Part II**
**Henry VI, Part III**
**Richard III**
**Comedy of Errors**
**Titus Andronicus**
**Taming of the Shrew**
**Two Gentlemen of Verona**
**Love's Labour's Lost**
**Romeo and Juliet**
**Richard II**

A Midsummer Night's Dream
King John
The Merchant of Venice
Henry IV, Part I
Love's Labour's Won
Henry IV, Part II
Henry V
Julius Caesar
Much Ado About Nothing
As You Like It
The Merry Wives of Windsor
Hamlet
Twelfth Night
Troilus and Cressida
All's Well That Ends Well
Measure for Measure
Othello
King Lear
Macbeth
Antony and Cleopatra
Coriolanus
Timon of Athens
Pericles, Prince of Tyre
Cymbeline
The Winter's Tale
The Tempest
Henry VIII

### Score a century

If you find you are really comfortable using the room system, you can expand it still further. All you do is add extra floors. For example, if you are able to easily picture ten items on one storey of your house, you can imagine a second floor with the same layout, and add another ten items. Keep going until you have a ten-storey skyscraper capable of holding 100 pieces of information in sequence. It may well be that you found it easy to store 20 items on one floor, in which case you only need five storeys to score a century of memories. You'll need to learn how to trigger the transfer to the next storey, probably by imagining stairs, or perhaps by deliberately chaining two images together at the point of transition.

## Root for routes

A logical development of the Roman room system is the journey method, in which you can choose any familiar route and store the items on it. This may suit people with a good sense of direction who are really good at finding their way about and never get lost once they've travelled a route. Keen geographers might even use a map and journey across a country or continent.

## Memory lane

Walk a route that you regularly travel – a stroll to the shops or a friend. You might prefer to imagine a car journey that you drive regularly. Select 20 locations where you will store memory images and learn these very thoroughly, in order, picturing the journey, reciting or sketching each place. You might choose traffic lights, a door, a shop front, a bridge (allowing you to look down from it to add a location), a hill (change of level can make the journey more memorable), train tracks, a bin, and so on. Test yourself until this is fully learned.

## Hanging around

Write a list of 20 everyday objects, or use the 20 items listed in the exercise on page 148. Learn the list in order using your route images. Make them as vivid as possible, with the items hanging or projecting from the location.

## Take me to the rivers

Use your list of 20 locations to learn the list of the longest rivers in the world (in order length).

1. Nile
2. Amazon
3. Yangtze
4. Mississippi/Missouri
5. Yenisei system
6. Obi
7. Yellow
8. Amur
9. Congo
10. Lena
11. Mackenzie
12. Peace
13. Finlay
14. Niger
15. Mekong
16. Rio de la Plata
17. Murray
18. Darling
19. Volga
20. Euphrates

This task will be much trickier than the previous one, where you could picture the items in some form on the route. This time you will probably need to use puns and wordplay, or to create associations using any other useful information. For example, put a pyramid next to the Nile and have a piranha fish snapping at you from the Amazon. We tend to identify paddle boat steamers and jazz bands with the Mississippi, you may have a 'yen' to see the Yenisei system (although it is in Mongolia and Russia, not Japan), and 'Obi' may make you think of Obi-Wan in the Star Wars films.

You can expand the journey method by adding locations and possibly by storing mnemonics at each place, so that your image triggers a set of data. One recent word memory champion stored three images at each location – tripling the amount of information he could recall with a minimum of extra effort.

Some people argue that the journey method is better suited to learning information in sequence than the room method, because you will always come across items in the same order on the journey, whereas sometimes you might scan a room in a different sequence – although if you train yourself to always work in the same direction this shouldn't be a problem. Practise seeing your journey backwards, which can be particularly useful for tracking back to identify a missing piece of information.

If you find the journey method works well for you, add more journeys to your memory bank. They can be real or imagined – you could just use a map. This should avoid any issues of images and information overlapping from different knowledge.

> ### Geological journey
>
> Try to learn the list of geological periods in order given on page 91. As the words may not mean much to you, you may need to use puns and wordplay or other triggers to create images that you can place at locations on your journey. So you are combining the journey method with other memorising techniques (as multi-sensory as you wish) to make some very tricky information memorable. There is also a mnemonic for this information on page 91 and you may find it helpful to place those words in order on your journey as your triggers, because these words carry more meaning.

You can extend your use of this method by recalling the journey backwards, or by commencing at different starting points. In this way you are manipulating the method and learning to control it.

**Why the journey method works**
This method is effective because we create more than one association for each piece of information, so there are more pathways to the memory. We remember the location itself but may well have other historical or sensory links to it, so there are more routes to find it. There are already a number of 'tags' tying that data with its partner information. The more vivid the images created at the location, the easier the brain finds it to recall them.

# REMEMBERING VERY LONG SEQUENCES

Many memory champions use this next technique, the major system, also called the phonetic number system. It has been around for about 300 years, so it has proved its value – but it takes a fair bit of work. The system is far more complex than other mnemonic systems, but once you have put the time and effort in and become accustomed to using it, it allows you to access a great deal of numerical information very efficiently.

## 300 years old and still going strong

The first version of the major system was introduced in 1648 by Stanislaus Mink von Wennsshein, (a pseudonym of Johann Just Winkelmann (1620–1699)). His system allocated letters to digits, but was refined over the years to free up the vowel sounds and make the method entirely phonetic. It is one of very few learning methods to have survived this long.

The underlying concept of the system is that we find it easier to remember words and images than numbers. It works by converting numbers into consonant sounds then 'filling in' these sounds with vowels to make words from which you can produce a memorable image.

# Major sounds

The consonant sounds for each number are:

| Number | Consonant sound | Reminders |
|---|---|---|
| 0 | s, soft c or z | Z is the first sound in zero (which also ends with a nought). |
| 1 | t, d or th | t and d only have one downstroke. |
| 2 | n | two downstrokes. |
| 3 | m | three downstrokes. |
| 4 | r | the last letter in four, and r looks like a reversed 4. |
| 5 | l | l is the Roman numeral for 50. |
| 6 | sh, ch, dg, soft g, j | j is a bit like a reversed 6. |
| 7 | k, hard c, hard g, ng | k could contain two 7s forming a sort of mirror image if you look very carefully! |
| 8 | v, f, ph, | An elaborate f looks like 8. |
| 9 | p or b | p is like a reversed 9. |

| | | | |
|---|---|---|---|
| 0 | zoo | 25 | nail |
| 1 | tie | | hinge |
| 2 | Noah | | neck |
| 3 | home | | knife |
| 4 | raw | | nap |
| 5 | hall | | |
| 6 | jaw | 30 | mouse |
| 7 | key | | mat |
| 8 | fee | | man |
| 9 | bee | | mime |
| | | | mayor |
| 10 | toes | | mail |
| | toad | | match |
| | tin | | mug |
| | dam | | movie |
| | door | | mob |
| | towel | | |
| | dish | 40 | rice |
| | deck | | rat |
| | dove | | rain |
| | tape | | ram |
| | | | rower |
| 20 | doze | | roll |
| | nut | | rage |
| | nun | | rock |
| | name | | roof |
| | near | | rope |

| 50 | lace | 75 | goal |
|---|---|---|---|
| | lid | | cage |
| | lion | | cake |
| | lamb | | cave |
| | lawyer | | cab |
| | lolly | | |
| | latch | 80 | face |
| | lock | | foot |
| | leaf | | fan |
| | leap | | foam |
| | | | fur |
| 60 | cheese | | file |
| | shed | | fudge |
| | chain | | fog |
| | jam | | five |
| | chair | | fob |
| | jail | | |
| | judge | 90 | bees |
| | check | | bat |
| | chef | | bone |
| | ship | | poem |
| | | | bear |
| 70 | case | | ball |
| | cat | | beach |
| | can | | book |
| | comb | | beef |
| | car | | pipe |

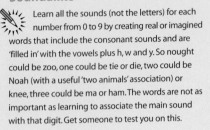

### Soundalike

Learn all the sounds (not the letters) for each number from 0 to 9 by creating real or imagined words that include the consonant sounds and are 'filled in' with the vowels plus h, w and y. So nought could be zoo, one could be tie or die, two could be Noah (with a useful 'two animals' association) or knee, three could be ma or ham. The words are not as important as learning to associate the main sound with that digit. Get someone to test you on this.

On page 132 we saw how the peg system can be expanded by choosing specific settings or images for each new tens number. In the phonetic system, this role falls to sounds, because two sounds can represent a pair of digits, enabling us to create a dictionary of words for all the numbers to 99. The numbers from ten to 19 start with a 't' or 'th' sound, while all the twenties numbers use an 'n' sound. So 20 must start with a 't' or 'd' sound for 'two' followed by a 'z' to show 'zero', so 'doze' does the job, while 88 requires two 'f' sounds, as in 'faff'.

The table has suggested words but you can, of course, make up your own following the same rules and, indeed, when you use the system you can change the word, provided the sounds used are consistent.

## Ten at a time

> Learn words to represent ten numbers, working a row at a time, until you can reliably identify a word to match any number up to 100. This should be done in stages.

## Major use

> Learn your pin number translated into the major system. Now learn your account number, which is much longer. You might also choose to learn words to represent door codes, passwords and other number sequences.

## Major dates

Equipped with a word for each number, you can create sounds for dates matching the association with the events identified with that date. For example 1066 could be rendered as '**t**o**e**s-**j**u**dg**e' – an image of a man in bare feet with a judge's wig on, perhaps standing watching poor old King Harold being blinded by an arrow. The year 1815 could be '**d**o**v**et**ow**el' – a white dove on a beach towel being thrown down by a defeated Napoleon at Waterloo.

## Make medical history

Use the major system to learn the dates of these ten events in medical history:

**1590** microscope invented
**1656** first intravenous medication
**1796** idea of vaccination introduced
**1888** first blood transfusion
**1895** X-rays first used
**1899** aspirin developed
**1922** insulin first injected
**1928** penicillin discovered
**1967** first heart transplant
**1983** HIV virus identified

## Other uses for the major system

The system is brilliant for number-related information such as birthdays and anniversaries. For example, 12 July (the seventh month) would be 'tin-key'. You could picture someone with that birthday holding a tin box with a key sticking out of it. And 15 November would be 'towel-toad' – you could picture a toad relaxing with the person after having a shower. It can also be used for any other sets of numbers, such as the 'chunked' elements of a credit card number.

## Remember your anniversary

Learn five anniversaries using the major system.

The major system is particularly useful for learning telephone numbers. For example, **06758 249314** is **zoo-check-leaf-near-poem-door**. You could imagine an animal cage with a tartan leaf floating past towards a sonnet stuck by a door. If you just enjoy the sounds created by the conjunction of the words, you could learn 'zuchikleevnrpomdor' which is easier to say than to write!

## Dial them up

Learn five telephone numbers using the major system. Test yourself after an hour, six hours, and a day.

The major system can help in numerous instances where we need to recall other number-related information, such as stocks and shares figures, page numbers, travel information, and remembering specialist terms in order. It is a mainstay of many memory champions.

While the major system can be used to create images for numbers, it can also of course be employed to number different objects in sequence, like a highly developed version of the peg system on page 132. Each number up to 99 is identified with a word that can be readily accommodated into an image, and is already associated with its neighbour. for example, if 60 is cheese and 61 is shed, the lump of cheddar for the first number is probably very close to the large item of garden furniture for the next. So you have the makings of a sequence. If you enjoy using the major system, it is worth seeing if it will allow you to remember very long sequences of information.

### 99 red balloons

Study the numbered list of 99 items. You will see that they are organised into themed groups, which should offer some help in this very challenging exercise. For each one, quickly create a clear image linking each with its major system image. You may want to pause every twenty items to give your brain time to reinforce those images before you go on. See if you can remember them in order.

| # | | # | | # | | # | |
|---|---|---|---|---|---|---|---|
| 1 | paperclip | 26 | carrot | 52 | wheelbarrow | 77 | make-up |
| 2 | pen | 27 | tomato | 53 | patio | 78 | actor |
| 3 | computer | 28 | onion | 54 | hedge | 79 | script |
| 4 | lamp | 29 | potato | 55 | tree | 80 | microphone |
| 5 | printer | 30 | tennis | 56 | flower | 81 | red |
| 6 | shelf | 31 | baseball | 57 | grass | 82 | blue |
| 7 | desk | 32 | cricket | 58 | rake | 83 | green |
| 8 | photograph | 33 | football | 59 | pot | 84 | yellow |
| 9 | file | 34 | hockey | 60 | jacket | 85 | white |
| 10 | cooker | 35 | rugby | 61 | tie | 86 | black |
| 11 | scales | 36 | badminton | 62 | trousers | 87 | orange |
| 12 | fridge | 37 | high jump | 63 | dress | 88 | purple |
| 13 | bin | 38 | javelin | 64 | skirt | 89 | brown |
| 14 | tap | 39 | swimming | 65 | socks | 90 | pink |
| 15 | toaster | 40 | hospital | 66 | vest | 91 | cake |
| 16 | coffee-maker | 41 | station | 67 | underwear | 92 | present |
| 17 | kettle | 42 | church | 68 | hat | 93 | invitation |
| 18 | knife | 43 | office | 69 | gloves | 94 | candles |
| 19 | fork | 44 | mosque | 70 | scarf | 95 | card |
| 20 | banana | 45 | car park | 71 | lights | 96 | song |
| 21 | orange | 46 | shop | 72 | camera | 97 | prize-bag |
| 22 | apple | 47 | house | 73 | action | 98 | magician |
| 23 | pear | 48 | skyscraper | 74 | film | 99 | balloon |
| 24 | kiwi fruit | 49 | airport | 75 | stage | | |
| 25 | pineapple | 50 | spade | 76 | screen | | |
| | | 51 | greenhouse | | | | |

Sprinting

## Memorising a deck of cards

You can adapt the major system to memorise a deck of cards. First, decide a word or sound to match each suit, using its initial letter. You could use the suit word itself, but it would be better to use its initial letter: hard C for clubs, D for diamonds, H for hearts, S for spades.

Combine these letter sounds with the words attached to each number, with aces counting as 1, jacks 11, queens 12 and kings 13. So the four of clubs needs a hard 'c' followed by 'r', making 'craw' or, if you prefer, 'car'. It doesn't matter that you could be using 'car' for 74, because in this context you know you are using the sounds in a different way. The four of spades starts with 's' to be followed by an 'r' sound, so the word could be 'sir' or 'sore'.

### House of cards

Create words for each card in a deck, working through one suit at a time. Do this slowly, with plenty of revision, working with one suit in each session until you are confident of remembering the words. Write out your dictionary of card words and get someone to test you on it.

Now you can try remembering a sequence of cards by combining this technique with the journey method, and creating an image from the card word at a location on a known route.

### Card order

Deal ten cards from a shuffled pack. Learn them in order by combining the card words into pairs. Repeat several times, testing for accuracy.

### Ace memory

Now try 20 cards. In addition to recalling them in sequence, practise picking out individual cards by where they are in the sequence, so that you can say 'The aces are the 12th and 17th cards'.

As you get better at this, increase the number of cards in the sequence until you can manage a whole deck. World memory champions aim to memorise ten decks.

You may be able to refine the technique by introducing the link system and putting a pair of images together at each location. So, for example the four of clubs followed by the four of spades would be 'car-sir' and you might image a hotel flunkey saying this as he hands over your keys.

## Remembering cards played in games

In many card games it is extremely valuable to know whether certain cards have already been played – and it doesn't matter when this happened, so you don't need to clutter up your memory with a sequence. This requires a different technique. You still use the card word, but when the card has been played, you imagine it mutilated in some way: broken, burned or cut. The image must be fairly dramatic to make it memorable.

### Kill the king

Deal out twelve cards from one suit, bring up your word image for each card dramatically disfiguring it in some way. What is the remaining card?

Repeat the 'Kill the king' exercise several times until you can readily alter the images of cards as they appear. Now we will try to identify two missing cards.

### Two missing

 Deal a shuffled pack of 24 cards taken from two suits, disfiguring the image of each card as it appears. Try to identify the two missing cards (which could, of course, both be from the same suit).

In the 'Two missing' exercise, see if you can tell whether a whole suit has appeared, because then you can use one disfigured image for that suit, rather than holding on to 13 manipulated images. When you are proficient at this, move on to a whole deck.

### Four gone

 Deal 51 cards from a whole shuffled deck. Can you identify the missing card? Once you can, deal one less card and pick out the two missing ones. Keep going until you can deal 48 from the deck and say which four cards remain.

# TOO DRY? DOMINIC TO THE RESCUE

Some memory enthusiasts find the major system a bit dry and dull because it is very hard to find vivid, memorable words for each number. The multi-memory championship winner Dominic O'Brien devised a variation on the system, called the Dominic system. Some consider it better because it links numbers and facts to names of people, so the images can involve celebrities or people you know in real life, or fictional characters.

The numbers 0 to 9 are given a letter as follows:

| Number | Letter |
|--------|--------|
| 0 | O |
| 1 | A |
| 2 | B |
| 3 | C |
| 4 | D |
| 5 | E |
| 6 | S |
| 7 | G |
| 8 | H |
| 9 | N |

Most of the letters are chosen to match their position in the alphabet, apart from 's' and 'n' which are both sounded twice in their numbers six and nine. This creates a pair of letters for each number up to 99. For example, ten is 'AO' and 99 is 'NN'. Single digit numbers have a zero put on the front, so one is 'OA'.

Next, you find a name that has the paired characters as its initial letters, so 72 (GB) could be Gordon Brown and 18 (AH) could be Adolf Hitler. The tricky bit is finding 100 names using the 10-strong alphabet used in the system.

### Mr Doyle, I presume

Once you have them, you can create images linking your cast of people with each other and with things you want to remember. For example, the first colour photograph was produced in 1872, so you could picture Adolf Hitler with his arm round Gordon Brown posing in front of a Victorian camera in brightly coloured clothes. To remember the pin number 6834 you could picture Sherlock Holmes (SH) tipping his deerstalker to thank an astonished Conan Doyle (CD) for inventing him.

> ### Gordon Brown is my bank manager
>
> Convert a set of ten four-figure numbers such as PIN codes and door codes into combinations of people. If you found this easy, make a note of the people you used and build a list of 100 names, using the letters in the Dominic system.

The system may well appeal to those who find it hard to conjure up images from a pair of nouns, but enjoy playing with the idea of putting people together in unlikely combinations. It is a fresh, fun approach that is just as effective as the major system and can, of course, be used to memorise a set of cards in a similar way.

The Dominic system can be linked with the journey method as you imagine a series of bizarre encounters with a blend of celebrities, acquaintances and fictional characters on the route of a journey. The nature of the journey – with its clear beginning, middle and end – turns the trip into a story and you may well find it useful to think in terms of a narrative rather than a journey, because it might be easier to add themes related to the facts you are trying to remember.

# THE STORY METHOD

## Once upon a time

The story method taps into our love of story – most of us can re-tell tales such as 'Red Riding Hood' many years after hearing them because they form so many strong pictorial, imaginative and emotional associations: a reminder to us that memory is multi-sensory.

The story method, in which you create a narrative around memory trigger images, can be used in various ways. In its simplest form, you can make up a tale using the very things you want to remember. So to recall a shopping list of apples, bananas, parsnips, dwarf beans, coffee, chocolates, flowers, beef, milk, bread, cheese, honey and jam you could make up a story in which Snow White takes the apple from the witch whose twisted nose looks like a parsnip and whose black hat reminds you of coffee. Realising she would be bananas to eat it, she calls to the seven dwarves (beans). They drive a stake (beef) into the witch, prepare a lunch of bread and cheese and go to live in the land of milk and honey, getting stuck in a traffic jam on the way. Then the prince arrives with chocolates and flowers.

### Once upon a time

Make up a simple story to remember a shopping list or similar set of information.

You can use images from the major method in a similar way. For example, to learn the credit card number **4791-6763-1289-3055** you use the words for the two-digit numbers: **rock-bat-check-judge-nail-five-mouse-lolly**: On a high rock lives a kindly bat. Every morning he checks that the local judge has woken up by rattling a tin full of nails outside his window surprising the mouse slurping on a lolly in the kitchen.

### Major story

Combine the story and major methods to remember a credit card or telephone number.

If you prefer the people-based Dominic method, the initials for the same number would be **DG-NO-SG-SC-AB-BS-CO-EE**. So the Director General of the BBC says 'NO' to actress Susan George's programme idea as they share a taxi with comic actor Steve Coogan and his unlikely new partner, movie legend

Anne Bancroft. They arrive at a gallery opening hosted by art critic Brian Sewell who welcomes them at the door and disdainfully points out movie idol Clive Owen being serenaded by Aussie housewife superstar Edna Everage.

> ### Celebrity figures
> Use the Dominic and story systems to create an unlikely celebrity tale to learn a credit or telephone number.

The story method is excellent for keeping information in order, so is ideal for sequences, However, it is pretty difficult to string together a story with more than 15 (or at the most 20) people or things – it tends to degenerate into a list, with the risk of some items being dropped. For longer sets of data, the link method (see page 119) is generally regarded as more reliable. As always, memory is a personal thing, and you will develop your own preferred methods and variations over time.

## Remembering a speech

A famous example of the value of the link/story method in public speaking is its use by author and lecturer Mark Twain. During his lifetime, he was probably as well regarded for his speech-making as for works such as *Huckleberry Finn*. Like any good public speaker, he knew that nothing turns an audience off as fast as reading out a verbatim speech: we talk better and more engagingly when apparently speaking 'off the cuff' without notes, as if we are making it up on the spot (although the phrase describes the method of writing brief reminders of the key elements of the talk somewhere inconspicuous).

### Get the picture

For example, a wriggly line under a haystack was the prompt to talk about ranch life in America's West. Next to it were slanting lines above an umbrella and the roman numeral II. This was the cue for a great wind that came at two o'clock every day.

He tried various techniques involving memorising about eleven key sentences, each leading to the next section of his lecture, but could not recall them in order. Twain switched to drawing images of what he wanted to say. He was no artist, but by planning the speech, identifying the prompts for new parts, he scribbled simple pictures, linking ideas. Twain found it best to then destroy the pictures, safe in the knowledge that they were embedded in his mind – and he claimed the images allowed him to recall whole speeches 25 years later.

### Talk of ten

 Plan a short talk on any subject you are interested in or need for your job. Identify the ten key transition points between each section and draw an image linking the two ideas, however crudely. You could keep these images to hand for your talk, or go the 'whole Twain' and destroy them. Now try your speech out on an audience.

### Story singing

If you remember tunes well, you might find it helpful to put your key ideas to a melody as well as or instead of drawing pictures. The mind is very good at picking up a tune and recalling how it develops, and you can exploit this by putting your own words to the melody.

### The tills are alive

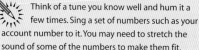 Think of a tune you know well and hum it a few times. Sing a set of numbers such as your account number to it. You may need to stretch the sound of some of the numbers to make them fit. Check your recall.

If that worked well, try this one:

> ### Tuneful facts
>
> Put together a few key words from a study pack or other information source. Set them to music. Check your recall of these words and their efficiency as memory triggers.

Bear in mind that the methods described in this section are powerful but require a great deal of preparation and practice: do not expect to win the world memory championships just yet. However, careful consideration of how they work for you should lead you to a better understanding of how you can boost your memory, perhaps by using elements of these techniques. Improving your memory improves your quality of life, but you'll have to do some spadework first!

# GLOSSARY

**Active reading:** A technique of reading and immediately categorising information for efficient memorising.

**Alzheimer's disease:** This is a neurodegenerative disease in which the memory, especially the short-term memory, deteriorates. It is the most common type of dementia.

**Amnesia:** Loss of short-term memory.

**Body system:** The imaginary placing of items on parts of the body to aid memorising.

**Chaining:** Creating a series of memories each triggering the next in the sequence (A triggers B, which triggers C, and so on).

**Chunking:** Grouping a series of words or numbers together for ease of learning.

**Concept map:** A way of linking images and information visually. A common format is to write the main subject in the centre of a piece of paper and draw lines away from it to create 'branches' for themed information.

**Declarative memory:** Where facts and experiences are stored. Also known as explicit memory.

**Dominic system:** Memory method devised by multi-world memory champion Dominic O'Brien. A variation on the major system, it links numbers and facts to names of people.

**Eidetic memory:** The ability to recall images, sounds, or objects in memory with great accuracy.

**Encoding:** The processing of taking in information.

**Episodic memory:** Where you store the biography of your life so far.

**Expanded rehearsal:** Method of learning information by returning to it, or summaries of it, at intervals.

**Flashbulb memory:** When people remember where they were when they heard dramatic news.

**Implicit memory:** See procedural memory.

**Grey matter:** Brain tissue.

**Journey method:** Memorising strategy in which you can choose any familiar route and store remembered items on it.

**Kinaesthetic learning:** Taking in information by touch or feel.

**Learning curve:** Term devised by Hermann Ebbinghaus to describe how we become more proficient at doing things as we become familiar with the task.

**Link system:** A method of remembering lists by creating associations between its elements in sequence.

**Loci system:** Memory system in which images are stored in sequence in a room or building.

**Long-term memory:** The permanent memory, divided into different types such as semantic, procedural and episodic memory.

**Major system:** Long-established memory system in which numbers are converted into sounds and words. Also known as the phonetic number system.

**Mnemonic:** General term for a memory aid such as a word, phrase or picture.

**Neuron:** Electrically active nerve cell in the brain.

**Neurotransmitter:** A chemical involved in the transmission of nerve impulses between nerve cells.

**Number rhyme system:** See peg system.

**Number shape system:** Memory method in which you give a number a pictorial mnemonic that shares its basic shape.

**Peg system:** Memory method in which each number up to ten is given a rhyming word partner which is paired with what you want to remember.

**Photographic memory:** The visual element of eidetic memory.

**Procedural memory:** Our knowledge of how to do things, such as riding a bike, which we do without consciously thinking about it. Also known as implicit memory.

**Recall:** Also known as retrieval, this is the finding of stored information.

**Reptilian brain:** The part of the brain stem which controls life functions and influences our fight, hide or run impulses.

**Semantic memory:** Where we store our memory of the world - the brain's equivalent of an encyclopedia.

**Short-term memory:** Also known as working memory, this is where we hold information before it is either forgotten or stored. It stores about seven items for 30-40 seconds.

**Storage:** The creation of a permanent record of encoded information.

**Story method:** Method in which a narrative is created using trigger images.

**White matter:** Glial cells that form a web of links crucial for an effective memory.

**Working memory:** See short-term memory.

## Answers

**Page 19 Where was it:** 1. Six. 2. Three. 3. Two. 4. Five. 5. Squares. 6. Black asterisk. 7. Blue circle. 8. Blue circle. 9. Black asterisk. 10. Black circle. 11. Black asterisk. 12. Black circle.

**Page 36 Backwards:** Reading a message backwards requires good memory and concentration.

**Page 36 Caesar Shift:** The ancient Romans used this code.

**Page 37 Pig Latin:** Pig Latin makes you process spoken information quickly.

**Page 56 Spot the difference:** (right).

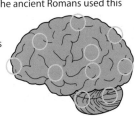

# FURTHER INFORMATION

## Websites

**www.aarp.org/nrta/Articles/a2003-08-19-memoryloss.html** a Washington, USA-based foundation that provides a variety of information for the over 50s.

**http://www.academictips.org/memory/index.html** offers guidance on using a variety of memory techniques.

**www.bbc.co.uk/radio4/memory/** has a series of interesting articles and discussions relating to a series of radio programmes on the subject.

**www.braingle.com/mind/** has a number of articles about memory and how to sharpen it.

**http://www.buzanworld.com/** website of memory expert Tony Buzan.

**http://www.ludism.org/mentat/MemoryTechnique** has detailed guidance on a range of memory techniques, including the Dominic system.

**www.mindtools.com/memory.html** offers advice geared to those taking exams or who need to remember detailed, structured information.

**www.nlm.nih.gov/medlineplus/memory.html** an American site which stores articles on medical research, including on memory.

**www.nongnu.org/majorteach/info/Mnemonics-concepts.html** shows how to use mnemonics.

**www.pseudonumerology.com/home.htm** has advice on how to create words to match long numbers.

**www.psywww.com/mtsite/norhyme.html** has information on a wide range of memory tools.

**stepanov.lk.net/mnemo/all3e.html** is a detailed look at modern mnemonic systems.

**www.wannalearn.com/Personal_Enrichment/Improve_Your_Memory/** has a number of instructional guides on memory techniques.

**www.web-us.com/memory/** offers a range of advice on memory techniques.

**www.worldmemorychampionships.com**, website of the world memory championships.

# Books

*Your Memory, a User's Guide*, Alan Baddeley (Penguin)

*The Handbook of Memory Disorders*, Alan Baddeley and others (Wiley and Sons)

*Tricks of the Mind*, Derren Brown (Transworld)

*Age-proof Your Brain: Sharpen Your Memory in 7 Days*, Tony Buzan (Harper Thorsons), one of numerous books on the subject by the same author.

*Mind Performance Hacks*, Ron Hale-Evans (O'Reilly)

*Your Memory*, Kenneth Higbee (Marlowe and Company)

*Memory Booster Workout*, Jo Iddon and Huw Williams (Hamlyn)

*Keep Your Brain Alive,* Lawrence Katz (Workman)

*Train Your Brain*, Ryuta Kawashima (Kumon Publishing)

*How to Develop a Brilliant Memory Week by Week*, Dominic O'Brien (Duncan Baird), one of many books by this multi-world memory championship winner.

*Brain Food,* Lorraine Perretta (Hamlyn)

*Stay Sharp with the Mind Doctor*, Ian Robertson (Vermilion)

*The Memory Bible: An Innovative Strategy for Keeping Your Brain Young,* Gary Small (Hyperion)

*Brain Power*, Marilyn vos Savant (Piatkus)

*Improving Your Memory*, David Thomas (Dorling Kindersley

# INDEX

acetylcholine 29, 31
acronyms 92–3
action triggers 115–16
advanced memory
techniques 141–82
alcohol 32
alphabet 95–7
system images
135–7
Alzheimer's disease 30,
33, 60
antioxidants 29
associated sequences 82
associations 81–4, 112,
128
astronomy 87
auditory memory 42,
43–6, 129
autobiographical
triggers 53–4

backwards recalling
20–21, 157–8
Blanes cells 60
'blind' activities 57–8
blueberries 29
body mass index (BMI)
32
body system 142–3
brain 4–6
fitness 14–74
size 12
waking up 75–112
brain stem 4
broccoli 29–30

Caesar shift 36
caffeine 32
calming down 27, 31
capital cities 120

carbohydrates 20
cards 34, 39, 41
memorising decks
134, 141, 170–71
memorising played
cards 172–3
categories 98–103
celebrities 178–9
cerebral hemispheres 6,
26
chaining (linking)
memories 119–26,
138–40, 179–80
change 34–5, 70–71
chess 48–9, 58
chunking 104–12
Cicero 144, 147
codes 36–7
colour 102–3
compass points 87
concentration 27
concept maps 140
consonants 160–64
continents 99
corpus callosum 6
counting 38–9
countries 95, 97, 99,
117–20
cramming 112
credit card numbers
107–27, 178
crosswords 33, 57
curry 30

dates 165–7
declarative memory 10
déjà vu 8
depressants 32
desks 72
detail recall 17–18

diet 29–32
distraction 73–4
Dominic system 174–6,
178–9
drugs 32

earworms 58–9
Ebbinghaus, Hermann
10–11
eidetic memory 48
emotional memory
64–5
encoding 7, 99
episodic memory 10, 16
escalator phenomenon
63
essential fatty acids 30
everyday activities,
brain-stimulating
33–5
everyday memory 20–21,
69–70
exams 112
expanded rehearsal
112, 124–5
extroverts 22

false associations 83–4
false memory syndrome
84
fats 30, 32
fibre 30
'fight-or-flight' reaction 4
fish 30
'Fizz buzz' game 39
flashbulb memory 65
free radicals 29
functional magnetic
resonance imaging
(fMRI) 13–14

games 34, 39–41
gender 12
geological time scales 91
ginkgo biloba 30
glial cells 4
grapheme colour synesthesia 103
grey matter 4

habits 70
herbs 31

introverts 22
IQ (Intelligence Quotient) 14
iron, dietary 30–31

jigsaws 33, 55
journey method 154–9, 170–71, 176
juggling 34

'Kim's game' 40
kinaesthetic memory 42, 89–90, 143

learning
 blocks to 28
language 115–16
 styles 42–49
lemon balm 31
life stories 16
linking memories 119–26, 138–40, 179–80
lists 71–2, 77
 and alphabetic systems 95–7
 categorising 98
 and the major system 168–9
 and the peg system 126–7

and route methods 145–9, 151–3, 155–8
shopping 71–2, 96, 98, 142–3, 178
Loci technique 144–54
long-term memory 9–10, 112
 auditory 45
 and expanded rehearsal 124–5
 visual 49

major system 160–74, 178
marketing 13–14
maths 37–9, 88
 *see also* numbers
meaningfulness 77
medical history 166
memory champions 112, 141, 160, 167, 171, 174
memory curves 10–11
memory triggers 66–74
 action triggers 115–16
 autobiographical 53–4
 objects as 70–71
 phrase triggers 80
 sounds as 116–17
mental pictures 82
 creation 114–18
 linking together 119–26, 138–40, 179–80
mnemonics 75–6, 86–94
 academic use of 87–9, 91
 creation 93–4

and the journey method 157–58
kinaesthetic 89–90
and song and rhyme 92, 94
and spelling 86–7
studying and 108–9, 110–12
months, days of 89–90
multi-sensory memory 53–65, 101, 177
music 89, 92, 94, 181–2

names 66–9, 100, 121–2
neurons 4
neurotransmitters 29
notes 71–2, 111–12, 140
nouns
 abstract 114
 concrete 115, 126–7
number phonetic system 160–74, 178
number rhymes 126–9, 130–34
 expanded 133, 134
number shape system 129–31
numbers 44, 47–8
 and associations 81–2
 chunking 105–8
 and patterns 78–80, 107

obesity 32
object triggers 70–71
O'Brien, Dominic 174
Omega 3 oils 30
ordered memories 82, 126–39, 142–73, 179

patterns 78–80, 107
peg system 126–9, 130–31

'Pelmanism' 41
personality type 22
phonetic number system 160–74, 178
photographic memory 48–9
phrase triggers 80
physical exercise 22–8
Pi 88
'Picnic game' 40
Pig Latin 37
PIN numbers 80, 165, 176
plaques, amyloid 30
Polish 116
post codes 93
PQRST 110–11
pregnancy 12
presentations 122–3, 179–81
procedural memory 10
prospective memory 10
puns 116–17, 157, 158
puzzles 33

rainbows 87
reading, active 99
recall 8, 11
  backwards 20–21, 157–8
  of detail 17–18
  everyday 20–21, 69–70
strategies 66–74
rehearsal, expanded 112, 124–5
retracing 73
rhyme 92, 94, 126–34
rivers 156–7
Roman room technique 144–54
rote learning 76

route methods 142–59, 170–71, 176
routine 34–5, 69–71, 70–71

sage 31
salad 31
Sambrook, John 127
self-awareness 42–3
semantic memory 9
senses 53–65, 101, 177
sequences, remembering 82, 126–39, 160–73, 179
Shakespeare, William 152–3
shopping lists 71–2, 96, 98, 143–3, 178
short-cut techniques 112–40
short-term memory 9, 51–2, 105
  auditory 44–5
  tests of 18
  visual 49, 55–6
sight 54–8
Simonides 147
skills 50
sleep 41
smell 60–61, 101
sound 59–60, 101, 129
  consonant 160–64
  rhyme 92, 94, 126–34
  triggers 116–17
  see also auditory memory; music; phonetic number system
spatial memory 19
speeches 122–3, 179–81

spelling 85, 86–7
spot-the-difference 55–6
SQ3R 110, 111
stimulants 32
storage 7–8
story method 177–82
story singing 181–2
strawberries 29
study methods 102–3, 108–12, 122–3
Su Doku 33

taste 64, 101
telephone numbers 105–7, 167
text messages 33, 62
tidiness 72
time chunking 109
tip of the tongue phenomenon 11–12
touch 62–3, 101
trans-fats 32
typing 61–2
Twain, Mark 179–80

untidiness 72

visual memory 19, 42–3, 46–9, 129–30, 140
voices 45

water consumption 29
white matter 4
word chunking 104–5
wordplay 116–17, 157, 158
working memory 9, 14, 51–2

yeast extract 31